Advances in 21st Century Human Settlements

Indexed by SCOPUS

This Series focuses on the entire spectrum of human settlements – from rural to urban, in different regions of the world, with questions such as: What factors cause and guide the process of change in human settlements from rural to urban in character, from hamlets and villages to towns, cities and megacities? Is this process different across time and space, how and why? Is there a future for rural life? Is it possible or not to have industrial development in rural settlements, and how? Why does 'urban shrinkage' occur? Are the rural areas urbanizing or is that urban areas are undergoing 'ruralisation' (in form of underserviced slums)? What are the challenges faced by 'mega urban regions', and how they can be/are being addressed? What drives economic dynamism in human settlements? Is the urban-based economic growth paradigm the only answer to the quest for sustainable development, or is there an urgent need to balance between economic growth on one hand and ecosystem restoration and conservation on the other – for the future sustainability of human habitats? How and what new technology is helping to achieve sustainable development in human settlements? What sort of changes in the current planning, management and governance of human settlements are needed to face the changing environment including the climate and increasing disaster risks? What is the uniqueness of the new 'socio-cultural spaces' that emerge in human settlements, and how they change over time? As rural settlements become urban, are the new 'urban spaces' resulting in the loss of rural life and 'socio-cultural spaces'? What is leading the preservation of rural 'socio-cultural spaces' within the urbanizing world, and how? What is the emerging nature of the rural-urban interface, and what factors influence it? What are the emerging perspectives that help understand the human-environment-culture complex through the study of human settlements and the related ecosystems, and how do they transform our understanding of cultural landscapes and 'waterscapes' in the 21st Century? What else is and/or likely to be new vis-à-vis human settlements – now and in the future? The Series, therefore, welcomes contributions with fresh cognitive perspectives to understand the new and emerging realities of the 21st Century human settlements. Such perspectives will include a multidisciplinary analysis, constituting of the demographic, spatio-economic, environmental, technological, and planning, management and governance lenses.

If you are interested in submitting a proposal for this series, please contact the Series Editor, or the Publishing Editor:

Bharat Dahiya (bharatdahiya@gmail.com) or

Loyola D'Silva (loyola.dsilva@springer.com)

Giuseppe T. Cirella · Bharat Dahiya
Editors

City Responses to Disruptions in 2020

From Lockdowns to Aftermath

Springer

Editors
Giuseppe T. Cirella
Faculty of Economics
University of Gdansk
Sopot, Poland

Bharat Dahiya
School of Global Studies
Research Center for Sustainable
Development and Innovation
Thammasat University
Bangkok, Thailand

School of Public Leadership
University of Stellenbosch
Stellenbosch, South Africa

ISSN 2198-2546 ISSN 2198-2554 (electronic)
Advances in 21st Century Human Settlements
ISBN 978-981-99-7987-5 ISBN 978-981-99-7988-2 (eBook)
https://doi.org/10.1007/978-981-99-7988-2

This Springer imprint is published by the registered company Springer Nature Singapore Pte Ltd.
The registered company address is: 152 Beach Road, #21-01/04 Gateway East, Singapore 189721, Singapore

Paper in this product is recyclable.

To countless people who lost their lives due to COVID-19

To millions of people who were impacted negatively by this deadly pandemic around the world

To the people who continue to suffer the aftereffects of COVID-19 and related treatments

Preface

City Responses to Disruptions in 2020: From Lockdowns to Aftermath aims to integrate the disciplines of economics and urban geography to create a framework of cooperation around the idea of urban economic stability. This global collaborative research project was borne in the midst of the COVID-19 pandemic in a discussion between the two of us in July 2021. We felt that it was important to provide a collaborative platform for scholars to share their rich research and insights on the disruptions caused by the pandemic.

The book explores these disciplines through the economic lens and creates a collaborative environment for addressing the global challenges caused by the COVID-19 pandemic. Environmental advocates and proponents of economic growth are increasingly at odds—having looked at the economic impact of the decline of the environment as well as the environmental loss that occurs with unchecked growth and urbanization. The pandemic changed the global scene; it shook the world's foundations, as a number of countries' lockdowns affected not only the global economy but also society and the environment. The global community has experienced the negative impact of the COVID-19 pandemic on our economies. There were steep declines in GDP, job losses reported in the millions, and many people experienced a decline in income.

An unplanned lockdown has taken its toll. Past shocks and how they have impacted urban economies as well as for how long are core to bettering our understanding of present and future urban economic change. The underlying economic factors that make a shock more damaging to certain economies or industries, as well as understanding the vulnerabilities of any given event, will help with the recovery process and allow for a better understanding of how negative impacts on individual businesses can be minimized.

The pandemic revealed the need to adopt a global development approach, taking into consideration four dimensions: global value chains, debt, digitalization, and the environment. Topics related to causation and lockdown are explored through a number of case studies from around the world. The chapters provide multifarious perspectives on how humanity has been affected as well as responded to the pandemic.

City and urban-related research from lockdowns to aftermath paints a picture of the disruptions in 2020 and our future ability to deal with the next crisis.

For contributing their innovative research, rich insights, and well-written chapters to this co-edited volume, we are grateful to all chapter authors: Godwin Arku, Tomasz Bieliński, Jana Chovancová, Sherin Jacob, Cengiz Kahraman, Eva Litavcová, Aurora Meliala, Matjaž P. Nekrep, Bruno E. Ongo Nkao, Maria de Lourdes Carneiro da Cunha Nóbrega, Igor Petruška, Anis Rifai, Ana Luisa Rolim, Alessio Russo, Mohammed Sharaf, Jacques S. Song, Janja Kramer Stajnko, Jesse Sutton, Isabella Leite Trindade, Agnieszka Ważna, Jonathan Andre Woods, and Vinita Yadav.

We would also like to thank Dr. Loyola D'Silva, Executive Editor, together with Springer Nature's entire team, for their support and cooperation throughout the preparation and publication of this co-edited volume.

Sopot, Poland — Giuseppe T. Cirella
Bangkok, Thailand, and Stellenbosch, South Africa — Bharat Dahiya

Contents

Editors and Contributors

About the Editors

Giuseppe T. Cirella Professor of Human Geography, works at the Faculty of Economics, University of Gdansk, Sopot, Poland, where he received a Doctor of Habilitation (*Dr. Hab.*) in Economics and Finance. He specializes in economic development, environmental social science, and sustainability. His interdisciplinary background also includes socio-political research throughout Eastern Europe, Africa, and China. After completing a Doctor of Philosophy (Ph.D.) at Griffith University, Australia, within the Centre for Infrastructure Engineering and Management, on developing a sustainability-based index, he founded the Polo Centre of Sustainability in Italy. Notably, he has held professorships and scientific positions at the University of Pretoria, South Africa; Saint Petersburg State University, Russia; Inner Mongolia University of Science and Technology, China; Life University, Cambodia; and Free University of Bozen, Italy. In his early career, he worked with the Canadian International Development Agency in Indonesia as well as with Radarsat International in Brazil. He is also on the editorial board of a number of internationally renowned academic peer-reviewed journals.

Bharat Dahiya is Director of the Research Center for Sustainable Development and Innovation at the School of Global Studies, Thammasat University, Thailand. He is an Extraordinary Professor at the School of Public Leadership, Stellenbosch University, Western Cape, South Africa.

An award-winning urbanist, Bharat combines cutting-edge research, policy analysis, and development practice aimed at examining and tackling socio-economic, environmental, and governance issues in the global urban context of sustainable development. Since the early 1990s, his research and professional work have focused on sustainable cities and urbanization, strategic urban planning and development, urban infrastructure, urban environment, climate change and urban resilience, post-disaster recovery and reconstruction, and cultural heritage and landscapes.

Working with the World Bank, UN-HABITAT, Asian Development Bank, UNDP, United Nations University, and Metropolis, he has initiated, led and/or contributed to international development projects and programs on sustainable urban development around the world. Some of the projects that Bharat led include the preparation of the first-ever *Citywide Pro-poor Ger-area Upgrading Strategy and Investment Plan for Ulaanbaatar, and Community-Led Ger Area Upgrading in Ulaanbaatar City* (Mongolia); 'Earthquake-resistant and Energy-efficient School Reconstruction in Sichuan Province', and 'Getting Children Back to School: Providing Prefabricated Classroom Units to the Earthquake-Affected Children of Gansu Province' (China); and 'City Development Strategy for Thanh Hoa City in a Regional Development Context' (Viet Nam).

Throughout his career, Bharat has been involved in the preparation and dissemination of cutting-edge knowledge products. At the World Bank headquarters in Washington DC, USA, he conducted the first-ever systematic review of the Bank's investments for improving urban liveability, published as a co-authored book, *Urban Environment and Infrastructure: Toward Livable Cities* (2004). At UN-HABITAT, he led, conceptualized, and coordinated the preparation of the United Nations' inaugural report on *The State of Asian Cities 2010/11* (2010). In 2018, he co-authored *Partnering for Sustainable Development: Guidelines for Multi-stakeholder Partnerships to Implement the 2030 Agenda in Asia and the Pacific* (UNU-IAS, and UN-ESCAP). He was the lead author of Metropolis's first-ever *Asian Metropolitan Report* (2021). He has co-edited three academic anthologies, *New Urban Agenda for Asia-Pacific: Governance for Sustainable and Inclusive Cities* (Springer, 2020), *Practising Cultural Geographies: Essays in Honour of Rana PB Singh* (Springer Nature, 2022), and *Disaster Resilience and Human Settlements: Emerging Perspectives in the Anthropocene* (Springer Nature, 2023).

Bharat advises several professional, academic, private sector, and non-profit organizations around the world. Since 2019, he has been a Member of the International Advisory Board for the UN-HABITAT's *World Cities Report* (Nairobi, Kenya). Since early 2014, he has been the Series Editor for the SCOPUS-indexed Springer book series, *Advances in 21st Century Human Settlements*, which has a growing collection of over 40 monographs and edited volumes. He serves as a member of the editorial board of several international journals as well as the International Scientific Board of the book series, 'Geographies of the Anthropocene', published by Il Sileno Edizioni, Italy.

He holds a Ph.D. in *Urban Governance, Planning and Environment* from the University of Cambridge, United Kingdom. He read for his M.A. in Geography from Jawaharlal Nehru University, and Master of Planning from School of Planning and Architecture, both based in New Delhi, India.

Contributors

Godwin Arku Department of Geography and Environment, Western University, London, ON, Canada

Tomasz Bieliński Faculty of Economics, University of Gdansk, Sopot, Poland

Jana Chovancová Faculty of Management and Business, University of Presov, Prešov, Slovakia

Giuseppe T. Cirella Faculty of Economics, University of Gdansk, Sopot, Poland; Faculty of Economics, University of Economics, Sopot, Poland

Bharat Dahiya School of Global Studies, Research Center for Sustainable Development and Innovation, Thammasat University, Bangkok, Thailand;
School of Public Leadership, University of Stellenbosch, Stellenbosch, Western Cape, South Africa

Maria de Lourdes Carneiro da Cunha Nóbrega Department of Architecture and Urbanism, Catholic University of Pernambuco, Recife, Brazil

Sherin Jacob St. Claret College, Bengaluru, India

Cengiz Kahraman Department of Engineering Sciences and Department of Maritime Transportation Management Engineering, Istanbul University-Cerrahpaşa, Istanbul, Turkey

Eva Litavcová Department of Mathematics, Faculty of Education, Catholic University in Ružomberok, Ružomberok, Slovakia

Aurora Meliala Faculty of Law, Universitas Pembangunan Nasional Veteran Jakarta, Depok, Indonesia

Matjaž P. Nekrep Faculty of Civil Engineering, Transportation Engineering and Architecture, University of Maribor, Maribor, Slovenia

Bruno E. Ongo Nkoa Center for Studies and Research in Economics and Management, University of Yaounde 2, Yaounde, Cameroon

Igor Petruška Faculty of Management and Business, University of Presov, Prešov, Slovakia

Anis Rifai Faculty of Law, Universitas Al-Azhar Indonesia, Jakarta, Indonesia

Ana Luisa Rolim Department of Architecture and Urbanism, Catholic University of Pernambuco, Recife, Brazil

Alessio Russo School of Architecture and Built Environment, Queensland University of Technology, Brisbane, Australia

Mohammed Sharaf Department of Engineering Sciences and Department of Maritime Transportation Management Engineering, Istanbul University-Cerrahpaşa, Istanbul, Turkey

Jacques S. Song Faculty of Economics and Management, University of Ebolowa, Ebolowa, Cameroon

Janja Kramer Stajnko Faculty of Civil Engineering, Transportation Engineering and Architecture, University of Maribor, Maribor, Slovenia

Jesse Sutton Department of Geography and Environment, Western University, London, ON, Canada

Isabella Leite Trindade School of Interior Design, Ryerson University, Toronto, ON, Canada

Agnieszka Ważna Faculty of Economics, University of Gdansk, Sopot, Poland

Jonathan Andre Woods Faculty of Law, Universitas Pembangunan Nasional Veteran Jakarta, Depok, Indonesia

Vinita Yadav Department of Regional Planning, School of Planning and Architecture, New Delhi, India

Urban Shocks from 2020: The Big Picture and Moving Forward

Giuseppe T. Cirella and Bharat Dahiya

Abstract Examining the multifaceted elements of urban economics, disruptions, and responses arising from the COVID-19 pandemic entails adopting a holistic perspective, encompassing the broader landscape, and exploring the diverse trajectories followed by regions worldwide in their recovery efforts. This chapter amalgamates economics and urban geography to establish a contextual framework that elucidates the stability of urban economies. Specifically, it highlights the significance of comprehending society-wide disturbances and their impact on urban economies, as well as the identification of vulnerabilities and economic factors that amplify such disturbances. The adoption of specific and case-related approaches is emphasized. A synopsis of the book "City Responses to Disruptions in 2020: From Lockdowns to Aftermath" delineates its chapters and illustrates the diverse effects of the COVID-19 pandemic, thereby creating distinct scenarios in relation to global value chains, debt, digitalization, and the environment across various cities and urban areas. The chapter places a strong emphasis on acquiring insights into the consequences of disruptions in 2020 and their implications for the future, with the aim of facilitating enhanced preparedness for forthcoming shock events in order to work towards achieving the United Nations' Sustainable Development Goal (SDG) 11: To make cities and human settlements inclusive, safe, resilient and sustainable. Drawing upon empirical data, statistical models, and case studies, this interdisciplinary and comprehensive approach offers actionable insights for policymakers, economists, and researchers.

Keywords Impact shocks · Global crisis · Urban economies · COVID-19 pandemic · SDG11 · Urban sustainability · Urbanization

---

G. T. Cirella (✉)
Faculty of Economics, University of Gdansk, 81-824 Sopot, Poland
e-mail: gt.cirella@ug.edu.pl

B. Dahiya
School of Global Studies, Research Center for Sustainable Development and Innovation, Thammasat University, Bangkok 10200, Thailand

School of Public Leadership, University of Stellenbosch, Stellenbosch 7602, Western Cape, South Africa

G. T. Cirella and B. Dahiya (eds.), *City Responses to Disruptions in 2020*, Advances in 21st Century Human Settlements, https://doi.org/10.1007/978-981-99-7988-2_1

1 Introduction

The premise of this chapter is to introduce the idea behind the "City Responses to Disruptions in 2020: From Lockdowns to Aftermath" book. It merges the fields of economics and urban geography, forming a cooperative framework centered around the concept of urban economic stability. The research in this book examines these disciplines from an economic perspective, fostering a collaborative environment to tackle the global challenges arising from the COVID-19 pandemic as well as future worldwide disruptions. Advocates for environmental preservation and proponents of economic growth find themselves increasingly at odds. They have examined the economic repercussions of environmental degradation, as well as the environmental losses associated with unregulated growth and urbanization (Kuddus et al. 2020; Armstrong et al. 2022; Seifollahi-Aghmiuni et al. 2022; Rakshit et al. 2023). The crisis of the COVID-19 pandemic dramatically transformed the global landscape (Hiscott et al. 2020; Leach et al. 2021).

The world shook to its core as various countries implemented lockdown measures, impacting not only the global economy but also society and the environment (Onyeaka et al. 2021; Rahman and Thill 2022; Lewis 2022). The international community witnessed the adverse effects of the pandemic on our economies, with significant declines in the gross domestic product, millions of job losses, and decreasing incomes for many individuals (Zhang et al. 2022; Yamaka et al. 2022). The unplanned shutdown also took a toll and acted as a shock to economies worldwide (Zhang et al. 2022; Yamaka et al. 2022; Cirella 2023). Understanding how previous shocks have affected urban economies, as well as their duration, is crucial for enhancing our comprehension of what happened in 2020 as well as how we might better plan for future urban economic shocks (Cirella 2023). Identifying the underlying economic factors that exacerbate the impact of shocks on specific economies and industries and comprehending these vulnerabilities, will aid in the recovery of such crises and provide valuable insights into the implementation of measures to mitigate the impacts on society. The pandemic has exposed the necessity for a case-by-case development approach that takes into account four dimensions: global value chains, debt, digitalization, and the environment (Hernández et al. 2014; Russo and Cirella 2017, 2020; Czermański et al. 2020; Fu et al. 2020; Czermanski et al. 2021; Loonam and O'Regan 2022). These dimensions are profoundly interconnected, forming a web of causation that reverberates globally, profoundly affecting economies, societies, and ecosystems (Cirella et al. 2020; Kahraman et al. 2022; Spring and Cirella 2022; Lewandowski and Cirella 2022; Fu et al. 2022). As countries grapple with the aftermath of the pandemic, it becomes evident that sustainable practices in global value chains are crucial for resilience and equitable growth (Kahraman et al. 2022). Similarly, debt management strategies that align with sustainability goals are essential to avert financial crises. Furthermore, the accelerated pace of digitalization during the pandemic necessitates a sustainable approach that mitigates its environmental impact. Examining these dimensions across diverse cities and conducting case studies from various regions of the world provides valuable insights into the dynamic interplay between human activities, sustainability, and resilience (Cirella and Zerbe 2014a; Cirella 2020, 2022; Oniszczuk-Jastrząbek et al. 2020; Jiao et al. 2021; Cirella et al. 2021). It is important that we draw from these experiences to shape a more sustainable and resilient future that safeguards both our globalized society and the fragile environment upon which it depends.

2 Navigating Impact Shocks: Analyzing Urban Economies and Fostering Resilience in Uncertain Times

A comprehensive examination of impact shocks is tested by utilizing theoretical frameworks on a range of cities and countries worldwide that have encountered declines in their urban economies. Through the lens of a sustainable growth and change model (Cirella et al. 2007; Cirella and Tao 2008, 2009a, b; Cirella and Zerbe 2014b; Cirella 2020), an evaluation of the repercussions of these shocks is needed to better understand and cope with future shocks. These concerns transcend the boundaries of economics and sociology, encompassing policymakers and ordinary individuals alike. The advent of the COVID-19 episode unleashed an unprecedented economic and social crisis, permeating all aspects of society and yielding detrimental consequences for consumers, investors, community development, and overall societal well-being (Ibn-Mohammed et al. 2021; Leach et al. 2021; Hosseinzadeh et al. 2022). If one delves into the lockdowns and the aftermath of this negative event, scrutinizing its impact on both demand and supply dynamics, the risks associated with moving forward in a business-as-usual manner is irresponsible and ill-advised (Ibn-Mohammed et al. 2021; Zhang et al. 2022; Hosseinzadeh et al. 2022; Yamaka et al. 2022). By examining various case studies from different parts of the world, a comprehensive understanding of the complexities and multifaceted nature of impact shocks on urban economies can be better understood. Case research can aid in unraveling the intricate web of causal factors, encompassing not only economic aspects but also social, environmental, and political dimensions. This approach delves into the intricate interplay between external shocks and internal vulnerabilities, highlighting the factors that amplify the impact of shocks on urban economies while shedding light on the mechanisms that could potentially mitigate such risks.

The research recognizes the importance of addressing the concerns of policymakers and everyday citizens. It endeavors to bridge the gap between theoretical frameworks and practical implications, providing actionable insights for policymakers to navigate the challenges posed by impact shocks and steer urban economies toward sustainable growth and resilience. Additionally, it aims to empower individuals by fostering a deeper understanding of the dynamics at play, equipping them with the knowledge necessary to navigate the uncertainties brought about by shock-induced disruptions (Cirella 2023). To accomplish these objectives, rigorous analytical work, drawing on a diverse array of quantitative and qualitative methods should be applied. As a result, this book leverages empirical data, statistical models, and comprehensive surveys to paint a comprehensive picture of the impact of shock-induced declination throughout the world. By employing this interdisciplinary approach, attempts are made to offer a holistic perspective that transcends traditional disciplinary boundaries. This book serves as a vital resource for policymakers and academicians seeking to comprehend and respond effectively to impact shocks on urban economies. Through its varied analyses, its integration of various disciplinary perspectives, and its emphasis on actionable insights, it strives to provide a roadmap toward achieving urban economic stability in the face of present and future global disruptions.

3 Synopsis of the Book

This book adopts a methodical approach, organizing its content into chapters composed of comprehensive case studies that revolve around specific cities and urban-related research from around the world. These case studies provide valuable insights into the multifaceted ramifications of the disruptions observed in 2020. It traces the enduring effects of these disruptions on current urban and economic trends, while also projecting their potential impacts on future trajectories. By thoroughly analyzing the research, the book strives to provide a comprehensive and holistic understanding of the broader implications and long-term consequences associated with the 2020 crisis. In all, it further aims at charting a course toward enhanced preparedness and resilience in anticipation of future shocks. The book presents concise case studies that effectively explicate the content covered in each chapter. A synopsis of the book chapters is presented in brief.

"The Impact of Policies on Cities' Resilience in Canada: A Multi-scalar Approach" (Sutton and Arku 2024) explores the 2020 crisis in terms of cities across Canada and their economic resilience (i.e., resistance and recovery). The research pieces together the degree of stringency of provincial policies, local-specific factors, and the industrial mix via a shift-share analysis. It was found that cities exhibited heterogenous resiliency to the crisis, with larger cities generally experiencing a greater capacity to recover. Overall, the results have important policy implications for local economic development, especially regarding the importance of local urban capacity.

"Why is Landscape Architecture Crucial for COVID-19 Recovery and Future Urban Shocks and Stresses?" (Russo 2024) serves as a wake-up call for policymakers and governments, highlighting the urgent need to regenerate cities in terms of the environment and urban green space. It is argued that from the post-pandemic perspective should create inclusive and resilient human settlements, ensuring that all people have access to high-quality green spaces. A literature review is conducted to explore the role of landscape architecture in terms of pandemic recovery, urban regeneration, and the building of resilience to future crises. It states that the pandemic presented an opportunity to reintegrate nature into the built environment and established more accessibility for green spaces, fostering healthy communities on a large scale.

"Transformation of Public Spaces Through Emergency Urbanism: Lessons from Toronto, Recife, and New York" (da Cunha Nóbrega et al. 2024) presents a comparative analysis of interventions involving the use of public spaces during the COVID-19 pandemic. It examines three major cities and identifies transformations that contributed to different regulations and urban plans. The research presents supporting evidence on how sustainable urbanism is essential to the balance of the ecosystem in the cities of Toronto, Recife, and New York. The interventions observed in each city are based on the conceptual foundations of sustainable urbanism (Farr 2007), research on the use of open spaces versus the detriment of closed spaces during the pandemic (Bazant and Bush 2021), recommendations and data from the World Health Organization (2021), and definitions on tactical urbanism and emergency urbanism. The conclusions veer toward how the city must deal with circulation, permanence, and commerce. It is argued that these three levels of action, acting together, generate conditions for the use of public space that strengthen the practice of sustainable urbanism—contemporarily.

"Informalized Workforce of Women Domestic Workers: Case of Bengaluru Metropolitan Region" (Yadav and Jacob 2024) researches women domestic workers from marginalized and underprivileged sections in Indian society. It was found that sustaining family life, due to the loss of income during the lockdown measures and strict enforcement during the COVID-19 pandemic, was problematic. An analysis was conducted to look at the precarious socioeconomic conditions faced by these women in Bengaluru. Key details included an examination of the struggles of workers, their relationship dynamics with employers, and the role of stakeholders in balancing the power relationship. Primary and secondary data was collected on socioeconomic conditions in pre- and post-lockdown. Strong evidence shows that during the pandemic the situation was much worse for women domestic workers than other social classes. It is recommended that capacity building in terms of human rights and the balancing of power dynamics be implemented.

"Critique of the Indonesian Omnibus Law Regime: Misguided Interpretation in Narrating 'Village Owned Enterprises'" (Meliala et al. 2024) delves into how the COVID-19 pandemic impacted the Indonesian legal frameworks in terms of economic law. As such, the Government of Indonesia proposed an acceleration to economic growth via the 'Village Owned Enterprise' (or *BUMDes*) regulation. This regulation has become a concern as it has deviated from the idea of building a constitutionally sound and economically prosperous form of democracy. While the constitution clearly mandates the development of Indonesia's economy through cooperativism, the government, instead, has focused on capital accumulation in the development of *BUMDes*. Research is developed on the subject matter in terms of the shift that occurred in villages around the country during the COVID-19 pandemic. It points toward how *BUMDes* act more like a limited company, i.e., controlled by the government, and interfere with the country's democratic principles and constitutional makeup.

"Environmental Footprint for the Accumulation of Plastic Waste During the COVID-19 Pandemic" (Kahraman et al. 2024) examines the environmental consequences of excessive use and improper disposal of plastics during the pandemic. Ecological impacts of plastic pollution during this period include single-use plastics and personal protective equipment. The increased production of medical waste and the prioritization of incinerating and landfilling plastic are also examined. Specific research looks at how the commencement of the vaccination campaigns and the generation of plastic-based medical and household waste during 2021 intensified (i.e., one year into the pandemic). Several recommendations for managing plastic waste, encompassing comprehensive and step-by-step approaches are described, i.e., sorting, sanitization activities, proper technological intervention, and process optimization. Furthermore, pivotal solutions such as the adoption of bioplastics and the substitution of fossil fuel-based plastics made of biodegradable or compostable materials are presented as alternatives.

"Urbanization and CO2 Emissions: Panel Data Analysis of EU countries" (Chovancová et al. 2024) investigates the dependence of CO2 emissions in terms of the urbanization level in 22 European Union countries. It explores their economic growth, total energy consumption, amount of energy produced from renewable sources, and nuclear energy as a low carbon energy source. Three key questions are addressed: (1) Are there long-term and short-term relationships between the variables examined? (2) What is the connection between carbon emissions and the independent variables at different periods? (3) Can urbanization, economic development, total energy consumption, and the share of renewable and low carbon energy sources predict the evolution of carbon emissions and thus the achievement of the EU's carbon neutrality targets? The research spans the period 1992–2019 using econometrics and general moments methods. Using a comparative sample from this period, can serve to assess the impacts of COVID-19 in terms of CO2 emissions during and after the pandemic as well as how the pandemic affected economies in the years following.

"E-scooter Sharing and COVID-19 Impacts on the Bike Sharing Market in Polish Cities" (Bieliński and Ważna 2024) explores nationwide case research of bike sharing and scooter sharing services in Poland. As a micro indicator in achieving the United Nations Sustainable Development Goals, particularly in terms of creating sustainable and livable cities, bike sharing systems in Poland were shut down during the 2020 lockdown while e-scooter sharing systems were not. This policy decision led to the fast development of e-scooter sharing which is perceived by users as safer than public transportation in terms of the threat of COVID-19. Most of the urban bike sharing systems recovered after the pandemic was declared over, but e-scooter systems remain dominant in the micro-mobility market around Poland. At the end of 2022, the e-scooter fleet was almost four times the size of the bike-sharing systems. This market transformation (and dominance) has created new vehicles, technologies, and changes in fleet management, and has left a long-term change throughout Poland's urban landscape.

"Impact of the Global Pandemic on the Activities and Preparedness of Protection, Rescue, and Relief Services in Slovenia" (Stajnko et al. 2024) focuses on the safety of the state and its citizenry against natural and other disasters. In Slovenia, the volunteer firefighting organizations are used as a case study to highlight their operational capability and 24-h standby and response time. The COVID-19 pandemic and the lockdown measures had a massive impact on their activities and preparedness in terms of protection, rescue, and relief forces. Since 2020 activities such as education and training have been disrupted. This disruption includes various social activities essential for building and sustaining the system. Recommendations are put forth to directly help with action plans to mitigate and restore the force's operational capacity to pre-pandemic levels.

"The Role of Cities and Urbanization in Africa: Temperature and Precipitation Analysis" (Nkao et al. 2024) examines how urbanization has affected the environment and climate on the African continent. The research investigates the effects of using a holistic approach known as the dynamic technique. Specifically, temperature and precipitation are measured to show a correlation between climate change and urbanization in urban areas around Africa. Using a sample of 52 African countries, the research estimates a panel data model using the ordinary least squares and the general least squares methods over the period 1991–2020. The preliminary results show that regardless of the dimension considered, urbanization increases climate change. The findings are robustly formulated by using urban population growth as an indicator of urbanization and sensitivity calculations to test agglomeration urban population, urban area, and urbanization rates.

4 Concluding Remarks

The year 2020 witnessed unprecedented disruptions in cities and urban areas around the world, after the COVID-19 pandemic was declared. In response to this crisis, cities implemented strict lockdown measures, imposing restrictions on mobility, economic activities, and social interactions. These lockdowns represented a significant deviation from the norm, as urban areas that once thrived on bustling streets and vibrant social gatherings came to a standstill (Hernández et al. 2014; Loonam and O'Regan 2022). The immediate aftermath of these lockdowns revealed the profound impacts on cities, with economic sectors such as hospitality, tourism, and retail experiencing severe downturns, resulting in job losses and financial instability for countless individuals and businesses (Hosseinzadeh et al. 2022; Lewis 2022). However, as the initial shocks subsided and cities gradually emerged from lockdown, a complex process of adaptation and transformation began to unfold.

Urbanization, a long-standing trend characterized by population growth, infrastructure development, and socioeconomic changes, faced new challenges and considerations. Cities had to rethink urban planning, transportation systems (Cirella et al. 2019), and public spaces to align with public health guidelines and mitigate the risk of the ongoing crisis. Concepts such as densification, urban resilience, and sustainable development took on renewed significance as cities sought to build and create more resilient urban environments. Moreover, the disruptions brought about by the

COVID-19 pandemic sparked a series of responses and transformations that have acted as a catalyst for rethinking urbanization processes, highlighting the need for resilience and inclusivity. As cities continue to grapple with the long-term impacts of the pandemic, the lessons learned from these disruptions can shape urban development strategies, fostering more equitable and responsive cities and urban solutions for the future. By synthesizing the empirical evidence presented in the case studies, it is hoped that the collective effort of building a more resilient and sustainable urban future will prioritize sound decision-making.

References

Armstrong JH, Nisi AC, Millard-Ball A (2022) A disciplinary divide in the framing of urbanization's environmental impacts. Conserv Sci Pract 4:e624. https://doi.org/10.1111/csp2.624

Bazant MZ, Bush JWM (2021) A guideline to limit indoor airborne transmission of COVID-19. Proc Natl Acad Sci 118:e2018995118. https://doi.org/10.1073/pnas.2018995118

Bieliński T, Ważna A (2024) E-scooter sharing and covid-19 impacts on the bike sharing market in polish cities. In: Cirella GT, Dahiya B (eds) City responses to disruptions in 2020: from lockdowns to aftermath. Springer Nature, Singapore

Chovancová J, Litavcová E, Cirella GT, Petruška I (2024) Urbanization and CO2 emissions: panel data analysis of EU countries (Chap. 8). In: Cirella GT, Dahiya B (eds) City responses to disruptions in 2020: from lockdowns to aftermath. Springer Nature, Singapore

Cirella GT (2020) Sustainable human-nature relations: environmental scholarship, economic evaluation, urban strategies. Springer, Singapore

Cirella GT (2022) Human settlements: urbanization, smart sector development, and future outlook. Springer, Singapore

Cirella GT (2023) Uncertainty shocks in Africa: impact and equilibrium strategies for sound economic and social development. Springer, Cham

Cirella GT, Tao L (2008) Measuring sustainability: an application using the index of sustainable functionality in South East Queensland, Australia. Int J Interdiscip Soc Sci Annu Rev 3. https://doi.org/10.18848/1833-1882/CGP/v03i08/52680

Cirella GT, Tao L (2009a) An adaptive quantitative method to measure sustainability: an application for the State of Queensland, Australia. Int J Environ Cult Econ Soc Sustain Annu Rev 5:127–139. https://doi.org/10.18848/1832-2077/CGP/v05i01/54563

Cirella GT, Tao L (2009b) The index of sustainable functionality: an application for measuring sustainability. World Acad Sci Eng Technol 3:268–274. https://doi.org/10.5281/zenodo.1330369

Cirella GT, Zerbe S (2014a) Quizzical societies: a closer look at sustainability and principles of unlocking its measurability. Int J Sci Soc 5:29–45. https://doi.org/10.18848/1836-6236/CGP/v05i03/51427

Cirella GT, Zerbe S (2014b) Index of sustainable functionality: application in Urat front banner. In: Cirella GT, Zerbe S (eds) Sustainable water management and wetland restoration in settlements of continental-arid Central Asia. Bozen University Press, Bozen, pp 137–155

Cirella GT, Tao L, Mohamed S (2007) An application of an adaptive quantitative method to measure the sustainability of the gold coast, Australia. J Coast Res SI50:52–56

Cirella GT, Bak M, Koźlak A, Pawlowska B, Borkowski P (2019) Transport innovations for elderly people. Res Transp Bus Manag 30:100381. https://doi.org/10.1016/j.rtbm.2019.100381

Cirella GT, Wanjiku S, Paczoski A, Tiruneh S (2020) Human-nature relations: the unwanted filibuster. In: Cirella GT (ed) Sustainable human-nature relations: environmental scholarship, economic evaluation, urban strategies. Springer, Singapore, pp 3–22

Cirella GT, Russo A, Benassi F, Czermanski E, Goncharuk A, Oniszczuk-Jastrząbek A (2021) Energy re-shift for an urbanizing world. Energies 14:5516. https://doi.org/10.3390/en14175516
da Cunha Nóbrega M de LC, Trindade IL, Rolim AL (2024) Transformation of public spaces through emergency urbanism: lessons from Toronto, Recife, and New York (Chap. 4). In: Cirella GT, Dahiya B (eds) City responses to disruptions in 2020: from lockdowns to aftermath. Springer Nature, Singapore
Czermański E, Pawłowska B, Oniszczuk-Jastrząbek A, Cirella GT (2020) Decarbonization of maritime transport: analysis of external costs. Front Energy Res 8
Czermanski E, Cirella GT, Oniszczuk-Jastrząbek A, Pawlowska B, Notteboom T (2021) An energy consumption approach to estimate air emission reductions in container shipping. Energies 14. https://doi.org/10.3390/en14020278
Farr D (2007) Sustainable urbanism: urban design with nature, 1st edn. Wiley, Hoboken, N.J.
Fu Y, Supriyadi A, Wang T, Wang L, Cirella GT (2020) Effects of regional innovation capability on the green technology efficiency of China's manufacturing industry: evidence from listed companies. Energies 13:5467. https://doi.org/10.3390/en13205467
Fu Y, Yang X, Wang T, Supriyadi A, Cirella GT (2022) Spatial pattern characteristics of the financial service industry: evidence from Nanjing, China. Appl Spat Anal Policy 15. https://doi.org/10.1007/s12061-021-09409-z
Hernández RA, Martínez Piva JM, Mulder N, United Nations (eds) (2014) Global value chains and world trade: prospects and challenges for Latin America. Economic Commission for Latin America and the Caribbean (ECLAC), Santiago, Chile
Hiscott J, Alexandridi M, Muscolini M, Tassone E, Palermo E, Soultsioti M, Zevini A (2020) The global impact of the coronavirus pandemic. Cytokine Growth Factor Rev 53:1–9. https://doi.org/10.1016/j.cytogfr.2020.05.010
Hosseinzadeh P, Zareipour M, Baljani E, Moradali MR (2022) Social consequences of the COVID-19 pandemic. a systematic review. Investig Educ En Enfermeria 40:e10. https://doi.org/10.17533/udea.iee.v40n1e10
Ibn-Mohammed T, Mustapha KB, Godsell J, Adamu Z, Babatunde KA, Akintade DD, Acquaye A, Fujii H, Ndiaye MM, Yamoah FA, Koh SCL (2021) A critical analysis of the impacts of COVID-19 on the global economy and ecosystems and opportunities for circular economy strategies. Resour Conserv Recycl 164:105169. https://doi.org/10.1016/j.resconrec.2020.105169
Jiao G, Lu L, Chen G, Huang Z, Cirella GT, Yang X (2021) Spatiotemporal characteristics and influencing factors of tourism revenue in the Yangtze river delta urban agglomeration region during 2001–2019. Sustainability 13:3658. https://doi.org/10.3390/su13073658
Kahraman C, Orobello C, Cirella GT (2022) Changing dynamics with COVID-19: future outlook. In: Cirella GT (ed) Human settlements: urbanization, smart sector development, and future outlook. Springer, Singapore, pp 235–252
Kahraman C, Sharaf M, Cirella GT (2024) Environmental footprint for the accumulation of plastic waste during the COVID-19 pandemic (Chap. 7). In: Cirella GT, Dahiya B (eds) City responses to disruptions in 2020: from lockdowns to aftermath. Springer Nature, Singapore
Kuddus MA, Tynan E, McBryde E (2020) Urbanization: a problem for the rich and the poor? Public Health Rev 41:1. https://doi.org/10.1186/s40985-019-0116-0
Leach M, MacGregor H, Scoones I, Wilkinson A (2021) Post-pandemic transformations: how and why COVID-19 requires us to rethink development. World Dev 138:105233. https://doi.org/10.1016/j.worlddev.2020.105233
Lewandowski R, Cirella GT (2022) Performance management systems: trade-off between implementation and strategy development. OperManag Res 16. https://doi.org/10.1007/s12063-022-00305-4
Lewis D (2022) What scientists have learnt from COVID lockdowns. Nature 609:236–239. https://doi.org/10.1038/d41586-022-02823-4
Loonam J, O'Regan N (2022) Global value chains and digital platforms: implications for strategy. Strateg Change 31:161–177. https://doi.org/10.1002/jsc.2485

Meliala A, Rifai A, Woods JA (2024) Critique of the Indonesian omnibus law regime: misguided interpretation in narrating 'village owned enterprises' (Chap. 6). In: Cirella GT, Dahiya B (eds) City responses to disruptions in 2020: from lockdowns to aftermath. Springer Nature, Singapore

Nkao BEO, Song JS, Cirella GT (2024) The role of cities and urbanization in Africa: temperature and precipitation analysis (Chap. 11). In: Cirella GT, Dahiya B (eds) City responses to disruptions in 2020: from lockdowns to aftermath. Springer Nature, Singapore

Oniszczuk-Jastrząbek A, Czermanski E, Cirella GT (2020) Sustainable supply chain of enterprises: value analysis. Sustainability 12:su12010419. https://doi.org/10.3390/su12010419

Onyeaka H, Anumudu CK, Al-Sharify ZT, Egele-Godswill E, Mbaegbu P (2021) COVID-19 pandemic: a review of the global lockdown and its far-reaching effects. Sci Prog 104:00368504211019854. https://doi.org/10.1177/00368504211019854

Rahman MdM, Thill J-C (2022) Associations between COVID-19 pandemic, lockdown measures and human mobility: longitudinal evidence from 86 countries. Int J Environ Res Public Health 19:7317. https://doi.org/10.3390/ijerph19127317

Rakshit B, Jain P, Sharma R, Bardhan S (2023) An empirical investigation of the effects of poverty and urbanization on environmental degradation: the case of sub-Saharan Africa. Environ Sci Pollut Res 30:51887–51905. https://doi.org/10.1007/s11356-023-25266-4

Russo A (2024) Why is landscape architecture crucial for COVID-19 recovery and future urban shocks and stresses? (Chap. 3). In: Cirella GT, Dahiya B (eds) City responses to disruptions in 2020: from lockdowns to aftermath. Springer Nature, Singapore

Russo A, Cirella GT (2017) Biophilic cities: planning for sustainable and smart urban environments. In: Aijaz R (ed) Smart cities movement in BRICS. Observer Research Foundation and Global Policy Journal, New Delhi, pp 153–159

Russo A, Cirella GT (2020) Urban sustainability: integrating ecology in city design and planning. In: Cirella GT (ed) Sustainable human-nature relations: environmental scholarship, economic evaluation, urban strategies. Springer, Singapore, pp 187–204

Seifollahi-Aghmiuni S, Kalantari Z, Egidi G, Gaburova L, Salvati L (2022) Urbanisation-driven land degradation and socioeconomic challenges in peri-urban areas: Insights from Southern Europe. Ambio 51:1446–1458. https://doi.org/10.1007/s13280-022-01701-7

Spring C, Cirella GT (2022) Fostering sustainable development: green energy policy in the European Union and the United States. Human settlements: urbanization, smart sector development, and future outlook. Springer, Singapore, pp 101–137

Stajnko JK, Nekrep MP, Cirella GT (2024) Impact of the global pandemic on the activities and preparedness of protection, rescue, and relief services in Slovenia (Chap. 10). In: Cirella GT, Dahiya B (eds) City responses to disruptions in 2020: from lockdowns to aftermath. Springer Nature, Singapore

Sutton J, Arku G (2024) The impact of policies on cities' resilience in Canada: a multi-scalar approach (Chap. 2). In: Cirella GT, Dahiya B (eds) City responses to disruptions in 2020: from lockdowns to aftermath. Springer Nature, Singapore

WHO (2021) WHO Coronavirus (COVID-19) Dashboard. World Health Organ. https://covid19.who.int. Accessed 16 Jul 2023

Yadav V, Jacob S (2024) Informalized workforce of women domestic workers: case of Bengaluru metropolitan region (Chap. 5). In: Cirella GT, Dahiya B (eds) City responses to disruptions in 2020: from lockdowns to aftermath. Springer Nature, Singapore

Yamaka W, Lomwanawong S, Magel D, Maneejuk P (2022) Analysis of the lockdown effects on the economy, environment, and COVID-19 spread: lesson learnt from a global pandemic in 2020. Int J Environ Res Public Health 19:12868. https://doi.org/10.3390/ijerph191912868

Zhang H, Li P, Zhang Z, Li W, Chen J, Song X, Shibasaki R, Yan J (2022) Epidemic versus economic performances of the COVID-19 lockdown: a big data driven analysis. Cities Lond Engl 120:103502. https://doi.org/10.1016/j.cities.2021.103502

The Impact of Policies on Cities' Resilience in Canada: A Multi-scalar Approach

Jesse Sutton and Godwin Arku

Abstract The 2020 economic crisis that stemmed from the COVID-19 pandemic has had an adverse effect on economies worldwide. The effect of the crisis has not been homogenous, with some cities being more adversely impacted than others. The economic crisis provides an opportunity to examine the multi-scalar dimension of localities' resilience. In particular, the economic crisis originates partly due to containment measures, which allows the examination of how national and subnational policies influence localities' resiliency. Specifically, this chapter examines the resilience (i.e., resistance and recovery) of Canadian cities to the economic crisis, investigating the multi-scalar influences of provincial policies to reduce the spread of the virus to determine if various degrees of policy stringency influenced cities' resilience. The research finds that the degree of stringency of provincial policies did not significantly affect the resiliency of cities. Additionally, when decomposing localities' resilience (i.e., resistance and recovery) into local-specific factors and industrial mix using the shift-share analysis, it is found that local-specific factors significantly determined the cities' resilience, whereas the industrial mix had a marginal effect. Further, cities exhibited heterogenous resiliency to the crisis, with larger cities generally experiencing a greater capacity to recover. Also, it appears that the crisis has had hysteretic effects on cities, with some cities exited with a greater post-growth level. In comparison, others exited with a lower post-growth level. Overall, the findings have important policy implications for local economic development, especially regarding the importance of local capacity.

Keywords Resilience · Multi-scalar · Local economies · COVID-19 pandemic · 2020 economic crisis

J. Sutton (✉) · G. Arku
Department of Geography and Environment, Western University, London, ON N6A 5C2, Canada
e-mail: jsutto22@uwo.ca

G. Arku
e-mail: garku@uwo.ca

G. T. Cirella and B. Dahiya (eds.), *City Responses to Disruptions in 2020*, Advances in 21st Century Human Settlements, https://doi.org/10.1007/978-981-99-7988-2_2

1 Introduction

The growth trajectory of cities is not a linear or smooth process but one periodically subjected to shocks. It is well understood that economies are inherently shock-prone (Martin 2018; Martin et al. 2015). Shocks have adverse and heterogeneous effects on local economies, such as increased unemployment and lowered long-term growth rates (Cerra and Saxena 2008; Martin and Sunley 2015; Sutton and Arku 2022a). The ability of localities to withstand and mitigate the effects of shocks has brought increased attention to the notion of resilience, especially since the 2008 recession (Sutton et al. 2023; Evenhuis 2017). The heterogeneous effects of shocks and localities' resilience to them are one of the central foci of economic geographers (see, for example, Boschma 2015; Bristow and Healy 2014, 2020; Evenhuis 2017; Martin and Sunley 2015; Simmie and Martin 2010).

Since 2010, resilience has become the new buzzword in the economic geography literature and in policy circles. It has gained substantial attention from academics, policymakers, and practitioners alike due to the real-world impacts that shocks have on cities' long-term growth trajectories and residents (Bristow and Healy 2020; Martin and Sunley 2015). Despite advancements in the resilience literature, resilience and its multidimensional and multi-scalar determinants are underexplored and require more in-depth empirical analysis (Bristow and Healy 2020). In particular, the multi-scalar determinants of resilience need to be investigated as it remains a 'black box' in the literature both theoretically and empirically.

The multi-scalar determinants of resilience are of mounting importance due to the increased global integration of economies over the past half-century (Dicken 2015). Local economies are not self-contained and autonomous entities but are shaped by their connections to and interaction with other economies at various spatial scales (Bristow and Healey 2020). As economies become increasingly integrated into the global economy, the spread of shocks is more rapidly dispersed and widespread (Streeck 2016; Zolli and Healey 2012). Despite the blurring of territorial economic boundaries due to globalization, local economies are not free from the constraints of the political, social, and economic landscape that they are embedded in. Cities operate within the confines of the larger power and political structure at the subnational and national levels to enhance their resilience, competitiveness, and sustainability (Cirella et al. 2007; Russo and Cirella 2020).

The 2020 economic disruption (which is still ongoing) caused by the 2019 COVID-19 pandemic provides a unique opportunity to investigate the contagious nature of shocks due to increased global economic integration and the role national and subnational policies play in influencing cities' resilience. Upon the COVID-19 outbreak from Wuhan, China, governments at various levels worldwide implemented stringent policies, known as containment measures (e.g., lockdowns and social distancing), to contain or curb the spread of the virus (Kang et al. 2020; Sattar et al. 2020). These policies had crippling effects on local economies, resulting in the disruption of supply chains and associated impact on demand, consumption, and overall economic activity (Bagchi et al. 2020; Gong et al. 2020). Attempts by governments to contain COVID-19 have had a rippling effect on the global economy, triggering a global economic crisis in 2020 (Borio 2020), and beyond. Upper-level government policies inhibited cities' adaptive capacity and hence, their resilience.

In Canada, this was exemplified by provincial policies (i.e., containment measures), which were marked by heterogeneity, imposing varying degrees of stringency on local economies within their administrative borders (Lawson et al. 2021; Cheung et al. 2021). Therefore, depending on their location in Canada, cities experienced heterogeneous effects during the 2020 economic crisis due to provincial containment measures. This raises the question: did the stringency of provincial policies influence the resilience of Canadian cities to the 2020 economic crisis caused by the COVID-19 pandemic?

This chapter aims to answer the question posed and, in doing so, fills the current gap in the resilience literature by investigating how provincial policies in Canada influenced the resilience of local economies to the 2020 economic crisis. Specifically, this chapter has three objectives. First, to assess cities' resilience (i.e., resistance and recovery) during the 2020 economic crisis. Second, to determine if the stringency of provincial containment measures by provinces influenced cities' resilience. Third, to decompose cities' resilience using shift-share analysis to assess what influenced local economic resilience. This chapter presents preliminary results of Canadian cities' resilience to the 2020 economic crisis as the pandemic is ongoing at the time of writing this chapter. Nonetheless, the chapter provides a novel examination of cities' resilience by situating them within the larger and multi-scalar construct of politics and policies in the context of the COVID-19 pandemic and the subsequent 2020 economic shock.

The remainder of this chapter is structured as follows. Section two provides a brief overview of the resilience literature. In addition, this section provides a discussion of the COVID-19 pandemic and its economic effect on Canada, locally and nationally, as well as the pandemics' effect on the global economy. Section three describes the methods used to conduct the analysis. Section four highlights the results. Section five provides a brief discussion and concludes in section six.

2 Background

2.1 Local Resilience, Determinants, and Shocks

Since its ascension in ecology (see, for example, Gunderson and Holling 2002; Holling 1973; Walker et al. 2004), the notion of resilience has rapidly spread across many disciplines. Resilience gained rapid popularity in economic geography and spatial economics after the 2008 financial recession (Bristow and Healy 2020), which amplified the perception in academia and political arenas that local, regional, and national economies are more susceptible and vulnerable to the increasing frequency of shocks due to globalization (Christopherson et al. 2010). Put simply, the current global economic landscape is marked by both increased permeability of localities and increased frequency of shocks as the local and global become progressively inseparable (Dicken 2015). Shocks in this context are analogous to the butterfly effect in which local disturbances can cause global crises. Furthermore, the notion gained popularity due to the effects that policies, such as BREXIT and austerity policies, have on localities when they are experiencing shocks (Martin 2012; Martin and Gardiner 2019).

Although there is no universally accepted definition of resilience (Martin and Sunley 2020), in the context of local economies, it generally refers to cities' ability to withstand, rebound from, and adjust to shocks. Scholars have identified four types of resilience: engineering, ecological, evolutionary, and transformative (Martin and Sunley 2020). Engineering resilience refers to the ability of and the speed at which local economies return to their pre-shock growth path when hit by a shock. Ecological resilience refers to the ability of economies to absorb a shock to maintain their growth path. However, when shocks exceed the threshold of economies' ability to absorb them, economies shift to a new, typically less ideal, growth path (i.e., equilibrium). Evolutionary resilience refers to economies' ability to maintain their core function and performance despite the shock by reorienting and reorganizing its structure to an existing or new and more favourable growth path (Boschma 2015; Martin and Sunley 2015; Simmie and Martin 2010). Transformative resilience refers to economies' ability to reconfigure and create a new set of structures and functions once environmental, social, or economic conditions make the current nature of the system untenable or undesirable (Banica et al. 2020; Martin and Sunley 2020).

To summarize, the first two types of resilience, brought forth by ecologist Holling (1973), interpret resilience as an economy maintaining or returning to an equilibrium state after a shock. Scholars criticize these interpretations of resilience for employing an equilibrium approach that emphasizes a stable equilibrium state. They argue that economies are never in a stable state but are in a constant state of uncertainty and change (Bristow and Healy 2014; Martin 2012; Simmie and Martin 2010). Additionally, the equilibrium approach cannot explain the unevenness of local resilience (Pike et al. 2010). The last two types of resilience focus on the adaptability, in partial or full capacity, of local economies, comprised of a collection of economic actors (i.e., entrepreneurs, firms, and institutions), whose behaviours determine the level of adaptability that occurs. Resilience, in this view, is a dynamic process, which is embedded in and part of the evolution and development of localities' economies (Boschma 2015; Martin 2012; Sutton and Arku 2022a).

The dynamic process of resilience consists of four dimensions: vulnerability, resistance, robustness, and recoverability (Fig. 1) (Martin and Sunley 2015). First, vulnerability refers to the sensitivity of cities' firms and their labour force to different economic shocks. Second, resistance refers to the impact that the shock has on local economies. Third, robustness refers to the ability of firms, institutions, and the labour force, which encompasses local economies to adapt and adjust to shocks (i.e., the localities' adaptive capacity), including multi-scalar political intervention and external agents and forces. Fourth, recoverability refers to the extent of post-shock

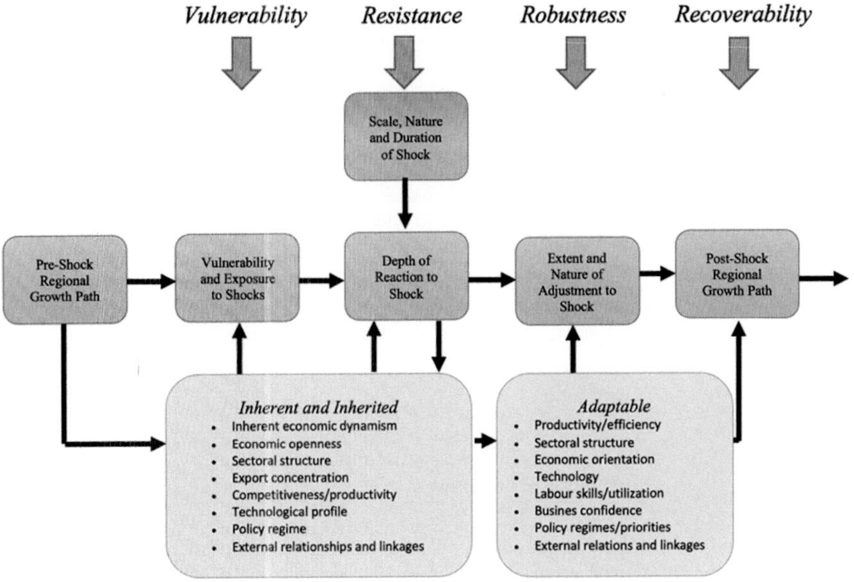

Fig. 1 Dimensions of resilience. *Source* Martin and Sunley (2015: 13)

recovery of local growth paths. These four dimensions are necessary to understand the resilience of local economies.

The determinants of local resilience are multidimensional and multi-scalar, comprised of internal (e.g., human capital, industrial structure, and social cohesion) and external factors (e.g., policies, knowledge pipelines, social networks, and trade flows). The external factors originate from various spatial scales (i.e., regionally, nationally, and globally). Specifically, local resilience is the product of internal and external key actors interacting with localities' innate structure and economic, social, and political landscape. During this interaction, key actors draw on localities' labour force and internal resources as well as introduce external resources. Interactions between key actors and localities are constrained by multi-scalar powers, politics, and institutions. Within these confines, the collective actions and decisions of the key actors determine localities' adaptive capacity and hence, overall resilience (Bristow and Healy 2018, 2020; Evenhuis 2017; Martin and Sunley 2015). The determinants of resilience are underdeveloped and underexplored, especially in recognizing the influence of policy and politics on local resilience. The importance of policies for resilience has been acknowledged (Bristow and Healey 2018), but empirical analysis on their influence has been relatively scarce (see, for example, Martin and Gardiner 2019; Sutton and Arku 2022b). Policies can either enhance or hinder localities' resilience when experiencing a shock (Martin 2012; Martin et al. 2016).

Shocks are sudden disturbances (Bristow and Healy 2020) that originate either endogenously or exogenously (Martin and Sunley 2015) and have the potential to

Table 1 Types of shocks

Types of shocks	Description	Example
Economic shocks	Disrupts the supply chain by reducing demand	2008 financial crisis
Institutional shocks	Institutional change that changes the economic landscape	BREXIT, COVID-19 policies, NAFTA, Austerity policy
Organizational shocks	Changes in industrial structure due to foreign competition alter global supply chains	Change in labour laws and consumption preference
Environmental shocks	Disrupts the supply chain by halting production	Earthquakes, floods, climate change, forest fires
Human-made shocks	Disrupts the supply chain by halting production	Terrorist attacks
Technological shocks	Disruptive technology that changes the economic landscape disrupting global supply changes	Steam engine and blockchain technology
Epidemic shocks	Alters the economic landscape and disrupts the supply chain by halting production	The COVID-19, SARS, and Ebola pandemics

disrupt the economic growth of localities (Banica et al. 2020). Shocks can be categorized into seven broad groups: economic, institutional, organizational, environmental, technology (Holm and Østergaard, 2015), human-made shocks, and epidemic shocks. Table 1 provides a description and example of each type of shock.

In response to shocks, local economies can be resistant (i.e., marginally impacted), resilient (i.e., severely impacted, but recovers), or non-resilient (i.e., severely impacted and does not recover) (Hill et al. 2012). Local economies' responses to shocks can shape their long-term growth trajectory as well as their resilience to future shocks (Martin and Sunley 2015; Simmie and Martin 2010). Additionally, the severity of shocks on localities depends on the origin, intensity, duration, and type of shocks experienced (Bristow and Healy 2020). Furthermore, shocks are not necessarily negative occurrences and can be positive events if cities, through their adaptive capacity, improve their long-term growth trajectory (Banica et al. 2020).

2.2 *COVID-19 Pandemic*

The first known case of COVID-19 was reported in Wuhan, China, in November 2019, which subsequently spread to over 200 countries and regions worldwide (Sattar et al. 2020). On March 11, 2020, the World Health Organization declared the COVID-19 outbreak a pandemic (WHO 2020). The pandemic manifested into a global shock that consisted of two major crises, a health and economic crisis, with the latter lagging behind the former (Gong et al. 2020; Kang et al. 2020). The measures implemented by countries to contain the spread of the virus caused the global economic crisis in 2020 (Borio 2020). This chapter focuses specifically on the latter.

2.2.1 Global Economic Impact

Pandemics, like COVID-19 in 2019 and SARS in 2002/2003, can have adverse and severe economic impacts on localities (Gong et al. 2020; Kang et al. 2020) due to the current integrated nature of the global economy. The COVID-19 pandemic has resulted in the largest global economic crisis since the Great Depression, surpassing the 2008 recession (Ibn-Mohammed et al. 2021). The pandemic has had heterogeneous impacts on industries, with manufacturing, tourism, retail, hospitality, and health industries being the most severely impacted (Gong et al. 2020; Sattar et al. 2020). In particular, the emergence of the 2020 economic crisis resulted from three intertwined effects of the pandemic (Gong et al. 2020). First, the outbreak resulted in workforce absenteeism in which workers who were infected with the virus had to take time off work. Workforce absenteeism results in a shortage of labour and reduced production. Second, government policies, such as lockdown measures, reduced the overall consumption and hence, the demand for goods and services. Third, the rising economic problems produced a ripple effect throughout global supply chains (Gong et al. 2020), such as interrupting the production and distribution of various commodities. In essence, the economic ramifications of the pandemic and government measures to curb the spread and contain the virus triggered an economic crisis manifested through the underlying mechanisms of globalization. The global economic impact is not surprising given that China has the second-largest economy and is the largest exporter of goods and services. Many of the world's largest companies are tightly intertwined with companies in China (Sattar et al. 2020).

Further complications arose due to the spatiotemporal effects of the virus. Countries were impacted by and responded to the virus at various times with measures of varying stringency, which episodically disrupted global supply chains to varying degrees throughout the pandemic. Additionally, countries are burdened with the tension borne from competing priorities. That is the tension between curbing or containing the spread of the virus through stringent policies and economic recovery entailing the removal or easing of said policies (Bagchi et al. 2020; Gong et al. 2020; Sattar et al. 2020; Sarkodie and Owusu 2021). Both options (i.e., maintaining the stringent policies or easing them) have ramifications of their own.

Although the pandemic is not over and subsequent waves could ensue, both the International Monetary Fund (2021) and the Organization for Economic Coordination and Development (2021) as of late 2021 state that the global economy is recovering and is forecasted to continue recovering into 2022. However, as mentioned above, successive waves of the COVID-19 pandemic could follow, resulting in further economic downturns in the global economy. The recovery of the global economy is tentative on the pandemic being contained.

Impact of the COVID-19 pandemic on Canada

Many countries have been impacted by the COVID-19 pandemic, and Canada is no expectation. The first reported case of COVID-19 in Canada was in Ontario on January 25, 2020, and Canada has since had over 4 million reported COVID-19 cases, with 3.7 million recoveries and 43,000 deaths (Goddard 2020; CTV 2022). The virus was initially brought into Canada by a man who visited Wuhan, China, and has since spread throughout the country due to community exposure (Global 2020). Due to the pandemic, provinces and territories have experienced health and economic impacts, albeit uneven (Statistics Canada 2022a; Soucy and Berry 2021). Figure 2 shows the economic impact of the pandemic on Canada's total employment. In 2020, Canada experienced a sharp decline in employment from 19.1 million in February to 16.7 million in June 2020. Since June 2020, Canada has experienced a steady increase in employment; however, with two subsequent minor shocks in January and June of 2021. Further, in November 2021, Canada fully recovered to its pre-crisis employment levels, indicating that Canada has recovered based on employment level.

In addition, the pandemic has also had heterogeneous economic impacts on Canadian cities (Table 2). Using the unemployment rate as an economic indicator, Table 2 illustrates the impact of the 2020 economic crisis on census metropolitan areas (CMAs) from 2019 to 2021—specifically focusing on the top five cities with the highest and lowest unemployment growth rates. As illustrated in Table 2, from 2019 to the onset of the pandemic in 2020, the unemployment rate increased for all CMAs to varying degrees, with Victoria (British Columbia) experiencing the highest rate of growth in unemployment of 1.28 and St. John's (Newfoundland and Labrador) experiencing the lowest at 0.27. Further, from 2020 to 2021, most CMAs have started to experience a decline in unemployment, with only two cities out of thirty-six (5.5%) continuing to experience increased growth rates in unemployment.

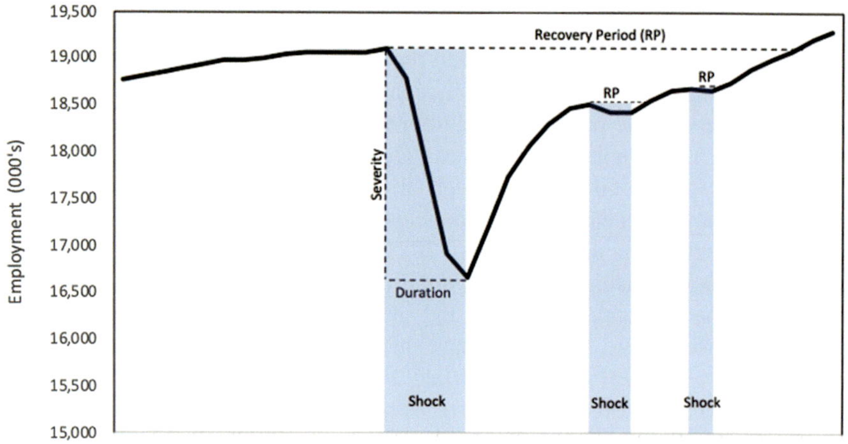

Fig. 2 Total employment in Canada, 2019–2020. *Source* Prepared by authors based on data obtained from Statistics Canada (2022c)

Table 2 Select Canadian cities' based on unemployment rate—2019 to 2021

Cities(n = 36)	Unemployment rate			Growth rate (2019–2020)	Growth rate (2020–2021)
	2019	2020	2021		
Top five growth rate (2019–2020)					
Victoria	3.4	7.7	5.0	1.28	−0.36
Québec City	3.1	6.7	4.3	1.17	−0.36
Guelph	4.5	9.2	7.1	1.04	−0.22
Hamilton	4.3	8.6	7.0	0.97	−0.19
Vancouver	4.7	9.1	7.2	0.95	−0.21
Bottom five growth rate (2019–2020)					
Abbotsford-mission	5.2	7.4	6.7	0.44	−0.10
Trois-Rivières	5.5	7.9	5.3	0.43	−0.34
Greater Sudbury	5.6	7.6	8.0	0.36	0.04
Moncton	5.8	7.7	7.7	0.32	0.01
St. John's	7.7	9.8	8.0	0.27	−0.18

Source Prepared by Authors based on data obtained from Statistics Canada (2022d)

To curb the spread of the virus, provincial governments in Canada imposed containment measures to varying degrees of stringency (Cheung et al. 2021). As of December 28, 2021, Canada has experienced five pandemic waves (i.e., spikes in infection rates), with containment measures tending to increase synchronously with each wave (see Sim 2021; CTV 2022). The impact of the provincial containment measures on the growth trajectory of cities within each province or territory is unknown. This chapter aims to examine the influence that policies have had on the resilience of Canadian cities.

3 Data and Methods

3.1 Data

The data used in this analysis is derived from two datasets: the Labour Force Survey (LFS), and the Bank of Canada. First, employment data is extracted from Statistics Canada's LFS using the National Industrial Classification System (two-digit level) for February 2016 to December 2021.[1] The LFS provides employment data at the disaggregated level of Census Metropolitan Areas (n = *36*) in Canada, which is the unit of analysis (Fig. 3). For context, it should be noted that Canada has a federal system consisting of ten provinces and three territories. There are no Census Metropolitan Areas in the Territories or in the Province of Prince Edward Island; therefore, only

[1] Statistics Canada (2022a, b).

Fig. 3 Census metropolitan areas, Canada. *Source* Prepared by authors based on data obtained from Statistics Canada (2022e)

nine provinces will be examined: Nova Scotia, New Brunswick, Newfoundland and Labrador, Quebec, Ontario, Manitoba, Saskatchewan, Alberta, and British Columbia. Second, stringency data was obtained from the Bank of Canada's stringency index for 2020 to 2021.[2] The stringency data indicates the restrictiveness of provincial containment measures and public information campaigns during the pandemic, with higher (lower) values indicating more (less) provincial stringent containment measures.

3.2 *Methods*

3.2.1 Measuring Local Resilience

Martin and Sunley (2015) note that there is no generally accepted method to operationalize resilience quantitatively. Thus, the analysis presented in the chapter follows Martin et al. (2016) and Martin and Gardiner (2019) to measure the resilience, specifically the resistance and recovery dimensions, of Canadian cities to the 2020 economic crisis. In particular, these authors use relative resilience (Eqs. 1–3), which uses the national resilience to shocks as an expectation or counterfactual in which cities are compared. Since the COVID-19 pandemic is a nationwide shock, a logical expectation is that ceteris paribus, cities, which compose the national economy, should react similarly to the macro-aggregate. To indicate the resilience of cities to shocks,

[2] Bank of Canada's Stringency Index: https://www.bankofcanada.ca/?p=219486.

employment is used instead of other indicators, such as income or gross domestic product. After a shock, similar output to pre-shock levels can occur without employment ever recovering. When a shock does occur, depending on the severity of the shock, it can cause a significant decline in employment, and severe and negative impacts on localities' labour market, resulting in increasing disparities among localities (Lagravinese 2015; Martin 2012). Therefore, this chapter uses Canadian cities' total labour force employment 15 years plus from 2016 to 2022 to measure relative resilience. The expected change of employment in city c during the shock or the recovery length of k years is expressed as:

$$(\Delta e_c^{t+k})^{\text{expected}} = \sum_i g_N^{t+k} e_{ic}^t \tag{1}$$

where: $(\Delta e_c^{t+k})^{\text{expected}}$ is the expected growth at time t to $t+k$ in city c given the national employment change; g_N^{t+k} is the nations' contraction (i.e., resistance) or expansion (i.e., recovery), in terms of local employment, to the COVID-19 pandemic; and e_{ci}^t equals employment in industry i in city c in starting time t (i.e., the base year). This equation produces a city's expected contraction and expansion during an economic shock. The measure of resistance (RES_c) is then given as:

$$RES_c = \frac{\left(\Delta e_c^{\text{contraction}}\right) - (\Delta e_c^{\text{contraction}})^{\text{expected}}}{\left|(\Delta e_c^{\text{contraction}})^{\text{expected}}\right|} \tag{2}$$

where: $\Delta e_c^{contraction}$ is the contraction of employment at time t to $t+k$ in city c. Although the pandemic is still ongoing, this chapter examines cities' short-term recovery (REC_c) similar to the measure of resistance. The measure of recoverability is given as:

$$REC_c = \frac{\left(\Delta e_c^{\text{expansion}}\right) - (\Delta e_c^{\text{expansion}})^{\text{expected}}}{\left|(\Delta e_c^{\text{expansion}})^{\text{expected}}\right|} \tag{3}$$

Equations 2 and 3 are centred around zero. Cities with a positive value were more resistant or recovered more than the national economy. In contrast, cities with a negative value were less resistant or recovered less than the national economy. For instance, a city under examination with a resistance and recovery value of 0.25 (−0.75) suggests that the city was 25% (−75%) more (less) resistant and experienced a 25% (−75%) greater (lower) recovery than the national economy. The resilience measures are used to examine the 2020 economic crisis, which impacted Canada in February 2020, with the nation starting to recover from June 2020.

3.2.2 Shift Share Analysis and Decomposing Resilience

The shift-share analysis is a robust analysis used in regional and urban analysis and has recently been employed, in various forms, to examine localities' resilience (e.g., Angulo et al. 2018; Martin et al. 2016; Ray et al. 2017; Tan et al. 2020). In the current analysis, the shift-share analysis is calculated using cities' employment data over the resistance and recovery period (i.e., RES_c and REC_c) of the 2020 economic crisis. The analysis decomposes cities' resistance and recovery into three effects: the national effect (NE), industrial mix (IM), and local effect (LE). The NE is the expected change in employment if cities' change in employment follows the national change in employment—which was used to calculate the expected values in Eqs. 1 and 2. The IM is the change in employment resulting from the growth in the industry, nationally. The LE is the change in employment due to local-specific (i.e., city-specific) factors or the city's competitiveness. The shift-share analysis determines the change in industry i in city c at time t to $t + k$ and is given as:

$$\Delta e_c^{t+k} = NE + IM + LE \tag{4}$$

$$NE = \sum_i g_N^{t+k} e_{ic}^t \tag{5}$$

$$IM = \sum_i (g_{iN}^{t+k} - g_N^{t+k}) e_{ic}^t \tag{6}$$

$$LE = \sum_i (g_{ic}^{t+k} - g_{iN}^{t+k}) e_{ic}^t \tag{7}$$

where g_N^{t+k}, mentioned above, is the national growth rate in employment over the resistance and recovery period; g_{ic}^{t+k} is the cities' growth rate in employment in a given industry over the specified period; and g_{iN}^{t+k} is the national growth rate in employment in a given industry over the specified period.

Following Tan et al. (2020), this study decomposes resilience (i.e., recovery and resistance) using the shift-share analysis. By subtracting the NE from both sides:

$$\Delta e_c^{t+k} - \sum_i g_N^{t+k} e_{ic}^t = \sum_i (g_{iN}^{t+k} - g_N^{t+k}) e_{ic}^t + \sum_i (g_{ic}^{t+k} - g_{iN}^{t+k}) e_{ic}^t \tag{8}$$

And then dividing both sides by NE:

$$\frac{\Delta e_c^{t+k} - \sum_i g_N^{t+k} e_{ic}^t}{|\sum_i g_N^{t+k} e_{ic}^t|} = \frac{\sum_i (g_{iN}^{t+k} - g_N^{t+k}) e_{ic}^t}{|\sum_i g_N^{t+k} e_{ic}^t|} + \frac{\sum_i (g_{ic}^{t+k} - g_{iN}^{t+k}) e_{ic}^t}{|\sum_i g_N^{t+k} e_{ic}^t|} \tag{9}$$

Equation (9) now equals cities' resistance or recovery as measured in Eqs. 2 and 3, with both dimensions of resilience being decomposed in its IM and RE. Under this restructured shift-share analysis, the IM indicates whether cities' resilience (i.e., resistance or recovery) is attributed to their industrial mix, which may be more or less resilient to the 2020 economic crisis compared to the national economy as a whole. Whereas, RE indicated whether cities' resilience (i.e., resistance or recovery) is due to its competitiveness, based on city-specific characteristics, such as local practitioners, policymakers, resources, human capital, and technological capacity.

4 Results

4.1 Local Resilience

Canadian cities demonstrated heterogenous resilience to the 2020 economic crisis. Figure 4 illustrates the resistance and recovery of Canadian cities to the crisis derived from Eqs. 1–3. As indicated by the top right quadrant of Fig. 4, no city was more resistant and had a greater recovery than the national economy. Twenty-seven cities were relatively less resistant to the economic crisis, but experienced greater recovery compared to the national economy, indicating that the majority of cities were severely impacted by the crisis. Also, the seven cities in the bottom left quadrant were the least resilient to the crisis, experiencing both lower resistance and recovery compared to the national economy. Four of the seven least resilient cities were located in the province of Ontario—the largest economy in Canada. Lastly, only two cities were relatively more resistant to the economic crisis than the national economy but experienced relatively lower recoveries. Furthermore, as illustrated in Fig. 4, there was a negative association between cities' resistance and recovery to the economic crisis, with a correlation of −0.67, which is moderately high, indicating that cities that were more resistant to the economic crisis tended to experience lower recovery.

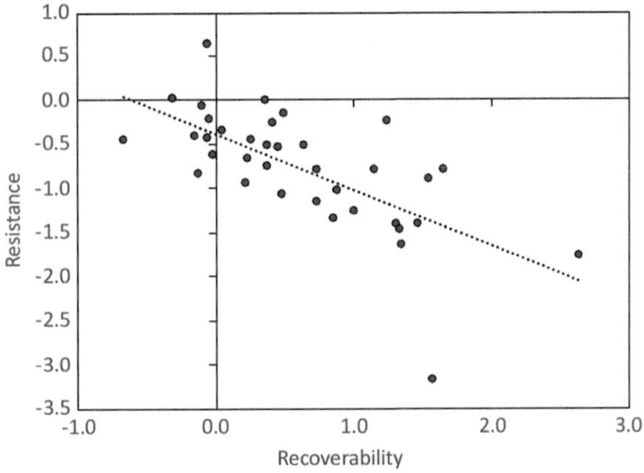

Fig. 4 2020 economic crisis, resistance (Feb 2019–Jun 2020) and recovery (Jun 2020–Dec 2020) of Canadian cities. *Source* Prepared by authors based on data obtained from Statistics Canada (2022a; b)

Table 3 Canadian CMAs' resistance and recovery to the 2020 economic crisis, Feb 2020–Dec 2021

CMA	Resistance	CMA	Recovery
Top five			
Moncton	0.62	Windsor	2.65
Kelowna	0.02	Peterborough	1.66
Saguenay	−0.02	Belleville	1.59
Sherbrooke	−0.07	London	1.55
Oshawa	−0.16	Edmonton	1.47
Bottom five			
Edmonton	−1.41	Sherbrooke	−0.08
St. Catharines-Niagara	−1.48	Trois-Rivières	−0.11
Vancouver	−1.65	Saint John	−0.14
Windsor	−1.78	Kelowna	−0.30
Belleville	−3.18	Barrie	−0.66

Source Prepared by authors based on data obtained from Statistics Canada (2022a, b)

Table 3 shows the cities with the top and bottom five resistance and recover to the crisis illustrating the varying degrees of resiliency. When examining cities resistance to the 2020 economic crisis, City of Belleville (Ontario), was the least resistant, with a value of −3.18, indicating that the city was 318% less resistant than the national economy. While the City of Moncton (New Brunswick), was more resistant, with a value of 0.62, suggesting that the city was 62% more resistant to the crisis compared to the national economy. Furthermore, evaluating cities' recovery from the 2020 economic crisis, City of Barrie (Ontario), recovered the least, with a value of −0.66, indicating that the city experienced 66% less recovery than the national economy. Whereas the City of Windsor (Ontario), recovered the most, with a value of 2.65, suggesting that the city experienced 265% greater recovery to the crisis compared to the national economy.

4.2 *Stringency Measures and Local Resilience*

Figure 5 shows the average stringency of provincial containment measures in 2020 and 2021. Provinces in the Prairie (Manitoba, Saskatchewan, Alberta, and British Columbia) and Atlantic (Nova Scotia, New Brunswick, and Newfoundland and Labrador (NFL)) regions tended to implement less stringent containment measures throughout 2020 when the pandemic first arrived in Canada compare to the other provinces with average values less than 44. Provinces of Ontario and Quebec implemented the most stringent containment measures throughout 2020, with average values of 51 and 49, respectively. Furthermore, throughout 2021 every province,

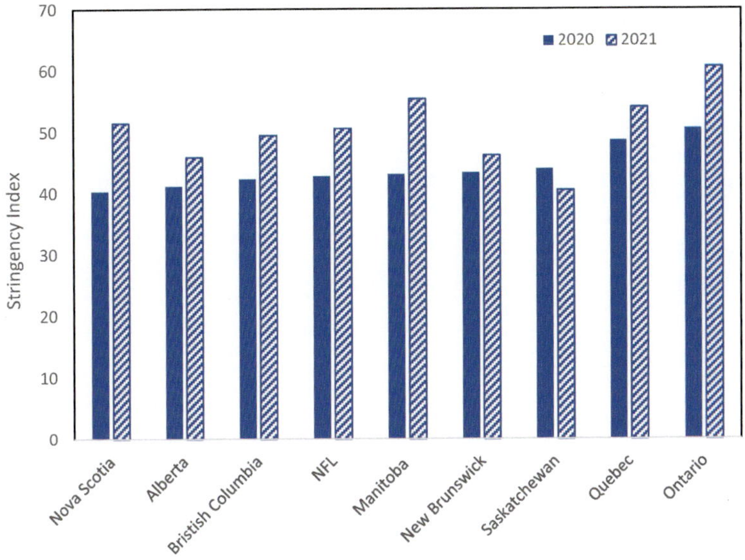

Fig. 5 Average stringency of provincial containment measures, 2020 and 2021. *Source* Prepared by authors based on data obtained from Cheung et al. (2021)

except Saskatchewan, on average implemented more stringent measures than they did in 2020. Overall, provinces implemented containment measures with varying stringency in 2020 (ranging from 40 to 51) and 2021 (ranging from 40 to 61).

To determine if the average stringency of provincial containment measures throughout 2020 and 2021 influenced cities' resilience (i.e., resistance and recovery), t-Tests and analysis of variances were conducted. Table 4 shows a two-sample t-Test assuming unequal variance for cities group by their natural clustering throughout 2020 and 2021. The results (i.e., t-critical two-tailed) indicate that the t-Test does not reject the research hypothesis that states that the two groups are equal. In other words, the results suggest that there is no difference between cities' resilience based on the stringency of provincial containment measures imposed on them. Several additional t-Tests and analyses of variances were conducted as a robustness check, such as assuming equal variance, dividing the natural clusters into three groups, and removing extreme data points; however, all results indicated did not reject the research hypothesis. Thus, the results indicate that provincial policies did not significantly influence cities' resilience capacity.

4.3 Decomposing Local Resilience

Figures 6 and 7 illustrate the decomposition of Canadian cities' resilience (i.e., resistance and recovery) into the industrial mix and local effect, based on Eq. 9. The

Table 4 Effect of stringency on cities' resilience, 2020 and 2021, two-sample t-test assuming unequal variance

	2020 stringency < 49	2020 stringency ≥ 49	2021 stringency < 54	2021 stringency ≥ 54
Mean	−0.59	−0.90	0.43	0.87
Variance	0.32	0.50	0.25	0.71
Observations	14	22	20	16
Hypothesized mean difference	0		0	
df	32		23	
t Stat	1.41		−1.84	
P(T ≤ t) one-tail	0.08		0.04	
t critical one-tail	1.69		1.71	
P(T ≤ t) two-tail	0.17		0.08	
t critical two-tail	2.04		2.07	

Source Prepared by authors based on data obtained from Cheung et al. (2021) and Statistics Canada (2022a, b)

results indicate that cities' resilience was significantly determined by city-specific characteristics (i.e., local effect). In contrast, the industrial mix had a marginal impact on cities' resilience and its effect, whether positive or negative, varied only slightly. Thus, the results indicate that the 2020 economic crisis negatively impacted most industries. In addition, larger core cities, like Toronto, Montreal, and Vancouver, were more severely impacted by the 2020 economic crisis. Yet, they experienced greater recovery than smaller peripheral cities, like Moncton, St. John's, and Lethbridge.

Furthermore, it is significant to note that the variation between cities' relative resistance and recovery indicates that local economies have undergone economic restructuring due to the economic crisis and subsequent recovery period (Figs. 6 and 7). City-specific characteristics, for example, human capital, resources, institutions, and technological capabilities, were essential for cities to recover and adjust their economy, resulting in some cities experiencing greater growth compared to other cities relative to the national economy. For instance, the City of Windsor was one of the least resistant cities to the economic crisis but experienced the greatest recovery; whereas, the City of Guelph was only marginally affected by the crisis compared to other cities and experienced substantial growth during the recovery period compared to other cities.

In summation, cities' resilience to the 2020 economic crisis depended on local-specific characteristics, with cities' industrial mix barely having any effect. Larger core cities experience the greatest fluctuation between the resistance and recovery period. Additionally, the 2020 economic crisis and cities' ability to recover has resulted in economic restructuring, suggesting that cities did not simply bounce back to their pre-crisis growth.

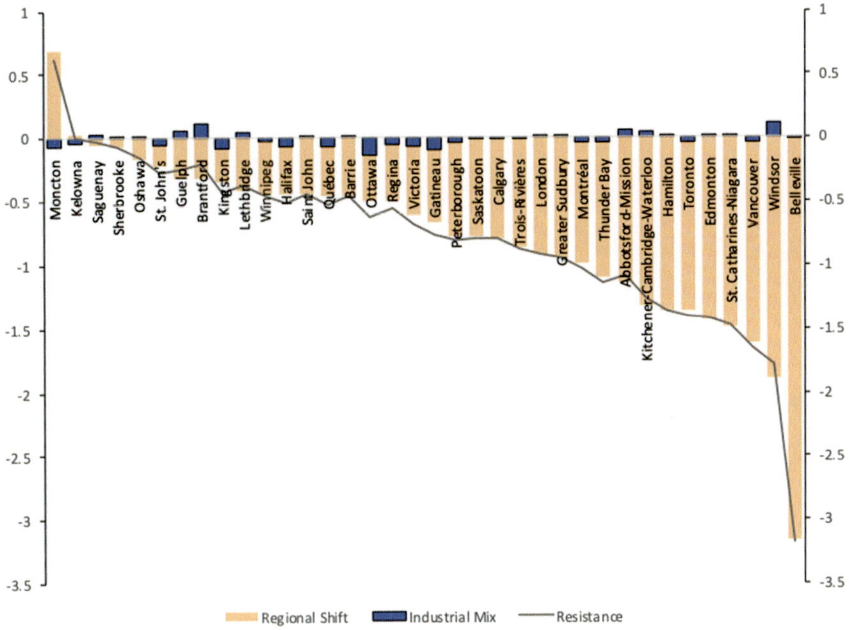

Fig. 6 Source of cities' resistance, 2020 economic crisis, Feb 2020–Jun 2020. *Source* Prepared by authors based on data obtained from Statistics Canada (2022a, b)

5 Discussion

This chapter set out to examine the influence of provincial containment measures and their stringency on Canadian cities' resilience to the 2020 economic crisis. In Canada, provinces implemented various degrees of stringent containment measures from 2020 to 2021. Although these policies impacted local economies, they did not statistically influence cities' resilience. That is, more or less stringent provincial policies did not influence their resistance and recovery to the 2020 economic crisis. These results contrast other studies that suggest national policies influence localities' resilience (e.g., Cuadrado-Roura and Maroto 2016; Webber et al. 2018). Therefore, policies, either national or subnational, may not influence localities' resilience to the degree indicated in previous findings.

While contrasting Cellini and Torrisi's (2014) findings, which indicate local economies display homogenous resiliency, the chapter's results confirm numerous empirical findings that local economies exhibit heterogeneous resilience to crisis (e.g., Fingleton et al. 2012; Sensier et al. 2016; Pontarollo and Serpieri 2020). In particular, Canadian cities' resilience, specifically resistance and recovery, to the 2020 economic crisis was marked by heterogeneity. This finding highlights the spatially heterogeneous capabilities and resources among Canadian cities that also

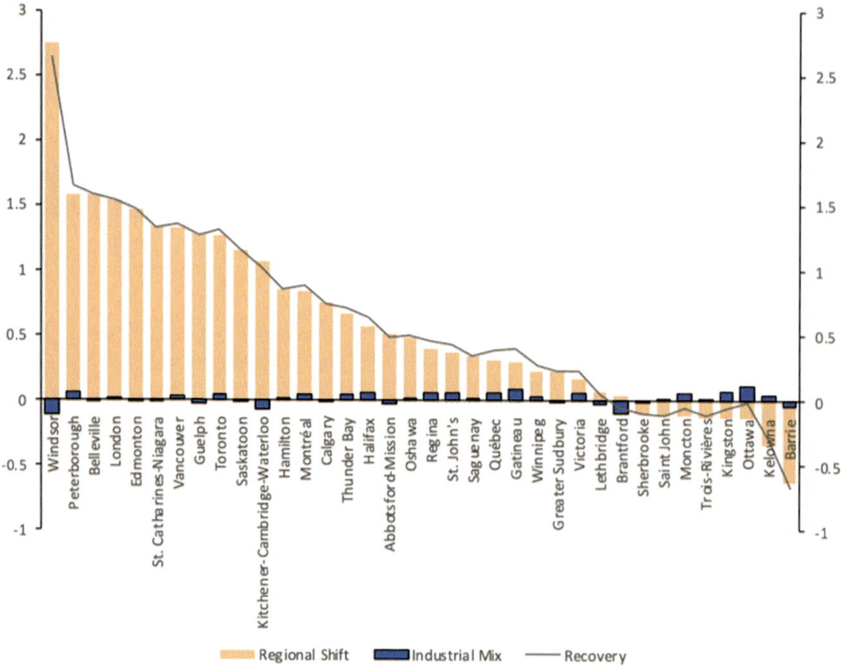

Fig. 7 Source of cities' recovery, 2020 economic crisis, Jun 2020–Dec 2021. *Source* Prepared by authors based on data obtained from Statistics Canada (2022a, b)

characterized localities in mature economies, such as Europe, the United States, and Australia.

This chapter found a negative association between cities' resistance and recovery to the 2020 economic crisis. This indicates that different determinants may enhance one dimension of localities' resilience (i.e., vulnerability, resistance, robustness, and recoverability) while hindering another, or that determinants may exhibit different functions depending on the dimension of resilience cities are in. For instance, diversification can enhance localities' resistance to shocks through a portfolio effect and enhance localities' reorientation (i.e., robustness) by providing more opportunities for recombination (Boschma 2015), highlighting the dual functions of diversification plays in localities' resilience. Also, Wolman et al. (2017) find that localities with a less educated labour force are less resistant to shocks but experience greater recoverability, highlighting that some determinants may enhance localities' resilience in one dimension but not another. Similarly, this chapter finds that larger core cities were less resistant to the crisis compared to smaller peripheral cities but experienced greater recovery, indicating the local-specific characteristics (e.g., proximity) may have hindered its ability to resist but enhanced its ability to recover. Therefore, the determinants of resilience should be broken down and discussed in the context of each dimension of resilience instead of being seen and discussed as a panacea for localities' resilience in its entirety.

While two cities (Moncton and Kelowna) were resistant to the 2020 economic crisis, the remaining cities were, to varying degrees, severely impacted. Some cities were more able to recover from the economic crisis than other cities. Conversely, some cities were less able to recover from the economic crisis than other cities. In both cases, cities experience varying impacts by the economic crisis; with some cities (e.g., Windsor and Guelph) recovering to a higher employment level after being severely impacted; while, others (e.g., Belleville and Ottawa) who were only marginally impacted by the economic crisis, did not rebound to their pre-crisis employment levels. The varying degrees of recoverability may indicate that Canadian cities underwent economic restructuring during the recovery period of the economic crisis. However, long-term analysis is needed to determine if some cities were slower to recover or if a permanent restructuring has occurred.

Furthermore, through decomposing resistance and recovery using the shift-share analysis, this chapter finds that cities' industrial mix did not significantly influence their resilience. In fact, all industries were affected by the 2020 economic crisis, though to slightly varying degrees. This finding indicates that cities' industrial base (i.e., diversification, specialization) may not always act as a determinant of resilience, and their effect may depend on the origin of the shock, such as shocks that are sector-specific, endogenous (e.g., major plant closure), or policy-induced (e.g., containment measures). In contrast, local-specific characteristics (i.e., local effects) determine cities' overall resilience. The influence of local factors emphasizes the importance of localities' characteristics, such as the skill of their labour force, entrepreneurial culture, institutions, social cohesion, resources, and so on. This finding contributes to the emerging evidence on the importance of inherent and inherited locally specific capabilities and resources for localities' resilience (Bristow and Healy 2017, 2018; Martin et al. 2016; Tan et al. 2020). Despite local economies becoming increasingly integrated into the global economy, local characteristics still matter, especially for resilience.

6 Conclusion

The 2020 economic crisis that followed the COVID-19 pandemic has provided a unique opportunity to examine the multi-scalar determinants of resilience by exploring the effects of national and subnational policies (e.g., provincial containment measures) on localities' resilience. This chapter has, in particular, investigated Canadian cities' resilience (i.e., resistance and recovery) to the 2020 economic crisis. It found that provincial containment measures did not significantly influence localities' resilience. Instead, local-specific characteristics determined their resilience both during the resistance and recovery periods. Therefore, the varying capacities of localities resulted in heterogenous resilience behaviour between cities. Also, cities' industrial mix did not affect their resilience, providing a stark contrast to much of the recent resilience literature that emphasizes the significance of industrial structure on determining localities resilience. However, the results should be interpreted with caution as the pandemic is still ongoing, and thus, subsequent waves and shocks could ensue.

While examining the multi-scalar determinants is a fruitful endeavour for future research, this chapter finds that there is still much work needed on the local determinants of resilience such as: what local-specific characteristics (e.g., social cohesion, institutions, and local policies) influence localities' resilience outside the dominant focus on industrial structure and human capital; what role do local practitioners and policymakers play in localities' resilience; and what effects do the determinants of resilience have on localities' resilience at each dimension of the resilience process? As data becomes more readily available on the 2020 economic crisis, literature should provide greater insight into the determinants of resilience as well as compare the effect of different types of shocks such as this epidemic shock and economic shocks.

The results from this chapter provide several policy implications. The analysis suggests that policies at all levels of government should focus on enhancing the local capacities to resist and recover from shocks, as local attributes primarily determine localities' resilience. In addition, practitioners and policymakers should avoid space-neutral policies as the results suggest that cities, depending on their size and location within provinces, demonstrate varying resilience capability, indicating that smaller peripheral cities may not have the resources and capability to recover from shocks as larger and core cities. Thus, place-specific policies should be implemented to address the disparities in localities' resilience capacity.

References

Angulo AM, Mur J, Trívez FJ (2018) Measuring resilience to economic shocks: an application to Spain. Ann Reg Sci 60(2):349–373

Bagchi B, Chatterjee S, Ghosh R, Dandapat D (2020) Coronavirus outbreak and the great lockdown. Springer Singapore Pte. Limited.

Banica A, Kourtit K, Nijkamp P (2020) Natural disasters as a development opportunity: a spatial economic resilience interpretation. Jahrb Reg 40(2):223–249

Borio C (2020) The Covid-19 economic crisis: dangerously unique. Bus Econ 55(4):181–190

Boschma R (2015) Towards an evolutionary perspective on regional resilience. Reg Stud 49(5):733–751

Bristow G, Healy A (2018) Economic crisis and the resilience of regions. Edward Elgar Publishing

Bristow G, Healy A (2014) Regional resilience: an agency perspective. Reg Stud 48(5):923–935

Bristow G, Healy A (2017) Innovation and regional economic resilience: an exploratory analysis. Ann Reg Sci 60(2):265–284

Bristow G, Healy A (2020) Handbook on regional economic resilience. Edward Elgar Publishing. https://doi.org/10.4337/9781785360862

Cellini R, Torrisi G (2014) Regional resilience in Italy: a very long-run analysis. Reg Stud 48(11):1779–1796

Cerra V, Saxena SC (2008) Growth dynamics: the myth of economic recovery. Am Econ Rev 98(1):439–457

Cheung C, Lyons J, Madsen B, Miller S, Sheikh S (2021) The Bank of Canada COVID-19 stringency index: measuring policy response across provinces. Bank of Canada. https://www.bankofcanada.ca/2021/02/staff-analytical-note-2021-1/
Christopherson S, Michie J, Tyler P (2010) Regional resilience: theoretical and empirical perspectives. Camb J Reg Econ Soc 3(1):3–10
Cirella GT, Tao L, Mohamed S (2007) An application of an adaptive quantitative method to measure the sustainability of the gold coast, Australia. J Coast Res SI50:52–56.
CTV News (CTV) (2022) Tracking every case of the COVID-19 in Canada. https://www.ctvnews.ca/health/coronavirus/tracking-every-case-of-covid-19-in-canada-1.4852102
Cuadrado-Roura JR, Maroto A (2016) Unbalanced regional resilience to the economic crisis in Spain: a tale of specialisation and productivity. Cambridge J Reg Econ Soc 9(1):153–178
Dicken P (2015) Global shift: mapping the changing contours of the world economy, 7th edn. Guilford Press
Evenhuis E (2017) New directions in researching regional economic resilience and adaptation. Geogr Compass 11(11):1–15
Fingleton B, Garretsen H, Martin R (2012) Recessionary shocks and regional employment: evidence on the resilience of UK regions. J Reg Sci 52(1):109–133
Global News (Global) (2020) Coronavirus: here's a timeline of COVID-19 cases in Canada. https://globalnews.ca/news/6627505/coronavirus-covid-canada-timeline/
Goddard E (2020) The impact of COVID-19 on food retail and food service in Canada: preliminary assessment. Can J Agric Econ 68(2):157–161
Gong H, Hassink R, Tan J, Huang D (2020) Regional resilience in times of a pandemic crisis: the case of COVID-19 in China. Tijdschr Econ Soc Geogr 111(3):497–512
Gunderson LH, Holling CS (2002) Panarchy: understanding transformations in human and natural systems. Island Press, Washington, DC
Hill E, Clair TS, Wial H, Wolman H, Atkins P, Blumenthal P, Ficenec S, Friedhoff A (2012) Economic shocks and regional economic resilience. In: Urban and regional policy and its effects. Brookings Institution Press, pp 193–274
Holling CS (1973) Resilience and stability of ecological systems. Annu Rev Ecol Syst 4(1):1–23
Holm JR, Østergaard CR (2015) Regional employment growth, shocks and regional industrial resilience: a quantitative analysis of the Danish ICT sector. Reg Stud 49(1):95–112
Ibn-Mohammed T, Mustapha K, Godsell J, Adamu Z, Babatunde K, Akintade D, Acquaye A, Fujii H, Ndiaye M, Yamoah F, Koh SC (2021) A critical analysis of the impacts of COVID-19 on the global economy and ecosystems and opportunities for circular economy strategies. Resour Conserv Recycl 164(1):105169–105169
International Monetary Fund (2021) World economic outlook: recovery during a pandemic, October 2021. https://www.imf.org/en/publications/weo
Kang M, Choi Y, Kim J, Lee KO, Lee S, Park IK, Park J, Seo I (2020) COVID-19 impact on city and region: what's next after lockdown? Int J Urban Sci 24(3):297–315
Lagravinese R (2015) Economic crisis and rising gaps north-south: evidence from the Italian regions. Camb J Reg Econ Soc 8(2):331–342
Lawson T, Nathans L, Goldenberg, Fimiani M, Boire-Schewab D (2021). COVID-19: emergency measures trackers, 28 October 2021. McCarthy Tetrault. https://www.mccarthy.ca/en/insights/articles/covid-19-emergency-measures-tracker
Martin R (2012) Regional economic resilience, hysteresis and recessionary shocks. J Econ Geogr 12(1):1–32. https://doi.org/10.1093/jeg/lbr019
Martin RL (2018) Shocking aspects of regional development: towards an economic geography of resilience. In: The new Oxford handbook of economic geography.
Martin R, Sunley P (2015) On the notion of regional economic resilience: conceptualization and explanation. J Econ Geogr 15(1):1–42
Martin R, Gardiner B (2019) The resilience of cities to economic shocks: a tale of four recessions (and the challenge of Brexit). Pap Reg Sci 98(4):1801–1832

Martin R, Sunley P (2020) Regional economic resilience: evolution and evaluation. In: Handbook on regional economic resilience. Edward Elgar Publishing, pp 10–35

Martin R, Sunley P, Tyler P (2015) Local growth evolutions: recession, resilience and recovery. Camb J Reg Econ Soc 8(2):141–148

Martin R, Sunley P, Gardiner B, Tyler P (2016) How regions react to recessions: resilience and the role of economic structure. Reg Stud 50(4):561–585

Organization of Economic Coordination and Development (OECD) (2021) OECD Economic outlook, interim report September 2021: keeping the recovery on track, September 2021. https://www.oecd.org/economic-outlook/

Pike A, Dawley S, Tomaney J (2010) Resilience, adaptation and adaptability. Camb J Reg Econ Soc 3(1):59–70

Pontarollo N, Serpieri C (2020) A composite policy tool to measure territorial resilience capacity. Socioecon Plann Sci 70:100669–100683

Ray DM, MacLachlan I, Lamarche R, Srinath KP (2017) Economic shock and regional resilience: continuity and change in Canada's regional employment structure, 1987–2012. Environ Plan A 49(4):952–973

Russo A, Cirella GT (2020) Urban sustainability: integrating ecology in city design and planning. In Cirella GT (ed) Sustainable human–nature relations: environmental scholarship, economic evaluation, urban strategies. Springer, pp 187–204. https://doi.org/10.1007/978-981-15-3049-4_10

Sarkodie SA, Owusu PA (2021) Global assessment of environment, health and economic impact of the novel coronavirus (COVID-19). Environ Dev Sustain 23(4):5005–5015

Sattar MF, Khanum S, Nawaz A, Ashfaq MM, Khan MA, Jawad M, Ullah W (2020) Covid-19 global, pandemic impact on world economy. Tech Soc Sci J 11:165–179

Sensier M, Bristow G, Healy A (2016) Measuring regional economic resilience across Europe: operationalizing a complex concept. Spat Econ Anal 11(2):128–151

Sim P (2021) Paisley sim: COVID-19 policy stringency across provinces. McGill University. https://www.mcgill.ca/maxbellschool/article/articles-max-policy/covid-19-policy-stringency-across-provinces

Simmie J, Martin R (2010) The economic resilience of regions: towards an evolutionary approach. Camb J Reg Econ Soc 3(1):27–43

Soucy JR, Berry I (2021) COVID-19 in Canada. University of Toronto, COVID-19 Canada Open Data Working Group. https://art-bd.shinyapps.io/covid19canada/

Statistics Canada (2022a) Employment by industry, monthly, seasonally adjusted and unadjusted, trend-cycle, last 5 months (x1,000). Table: 14-10-0355-01 (formerly CANISM 282-0088). https://www150.statcan.gc.ca/t1/tbl1/en/tv.action?pid=1410035501

Statistics Canada (2022b) Employment by industry, three-month moving average, unadjusted for seasonality (x1,000). Table: 14-10-0379-01. https://www150.statcan.gc.ca/t1/tbl1/en/tv.action?pid=1410037901

Statistics Canada (2022c) Labour force characteristics, three-month moving average, seasonally adjusted. Table: 14-10-0380-01. https://www150.statcan.gc.ca/t1/tbl1/en/tv.action?pid=1410038001

Statistics Canada (2022d) Labour force characteristics, three-month moving average, unadjusted for seasonality. Table: 14-10-0378-01. https://www150.statcan.gc.ca/t1/tbl1/en/tv.action?pid=1410037801

Statistics Canada (2022e) 2021 Census – boundary files. https://www12.statcan.gc.ca/census-recensement/2021/geo/sip-pis/boundary-limites/index2021-eng.cfm?year=21

Streeck W (2016) How will capitalism end? Verso, London

Sutton J, Arku G (2022a) Regional economic resilience: towards a system approach. Reg Stud Reg Sci 9(1):497–512

Sutton J, Arku G (2022b) The importance of local characteristics: an examination of Canadian cities' resilience during the 2020 economic crisis. Can Geogr 66(4):712–727

Sutton J, Arcidiacono A, Torrisi G, Arku RN (2023) Regional economic resilience: a scoping review. Prog Hum Geogr 47(4): 500–532
Tan J, Hu X, Hassink R, Ni J (2020) Industrial structure or agency: what affects regional economic resilience? Evidence from resource-based cities in China. Cities 106:1–11
Walker B, Holling CS, Carpenter SR, Kinzig A (2004) Resilience, adaptability and transformability in social–ecological systems. Ecol Soc 9(2):5
Webber HA, Bristow G (2018) Regional growth paths and resilience: a European analysis. Econ Geogr 94(4):355–375
Wolman H, Wial H, Clair TS, Hill E (2017) Coping with adversity. Cornell University Press. https://doi.org/10.7591/j.ctt1w0dc2v
World Health Organization (WHO) (2020) WHO Director-General's opening remarks at the media briefing on COVID-19 - March 11 2020. https://www.who.int/director-general/speeches/detail/who-director-general-s-opening-remarks-at-the-media-briefing-on-covid-19---11-march-2020
Zolli A, Healy AM (2012) Resilience: why things bounce back (1st Free Press hardcover ed.). Free Press

Why is Landscape Architecture Crucial for COVID-19 Recovery and Future Urban Shocks and Stresses?

Alessio Russo

Abstract The recent pandemic has served as a wake-up call for policymakers and governments, highlighting the urgent need to regenerate our cities in preparation for future environmental shocks. Studies have indicated that the risk of COVID-19 transmission was significantly lower in urban green spaces, providing a safe and much-needed refuge for many individuals living in lockdown. Thus, in a post-pandemic world, it is essential to create inclusive and resilient human settlements, ensuring that everyone has access to high-quality green spaces. The provision of such green spaces not only enhances citizens' quality of life but also improves human health and well-being while simultaneously reducing healthcare costs. Therefore, prioritising the opportunity to cultivate more food within our cities should be recognised as a public health priority. Within this chapter, a literature review is conducted to explore the role of landscape architecture in pandemic recovery, urban regeneration, and building resilience to future shocks. The pandemic has presented us with an opportunity to reintegrate nature into our built environment and establish more accessible green spaces. The findings from the literature demonstrate that landscape architecture will play a fundamental role in driving a green recovery and fostering the development of healthy communities.

Keywords Pandemic · Climate change · Green recovery · Green infrastructure · Urban resilience

1 Introduction

In a post-pandemic world, it is evident that the COVID-19 pandemic has had a profound impact on the health of urban dwellers, resulting in widespread devastation. The infection has claimed the lives of hundreds of thousands of individuals,

A. Russo (✉)
School of Architecture and Built Environment, Queensland University of Technology, Brisbane, Australia
e-mail: alessio.russo@qut.edu.au

G. T. Cirella and B. Dahiya (eds.), *City Responses to Disruptions in 2020*, Advances in 21st Century Human Settlements, https://doi.org/10.1007/978-981-99-7988-2_3

underscoring the severity of its consequences. Concurrently, public health and societal initiatives implemented to combat the transmission of the virus continue to yield far-reaching health repercussions (WHO 2022). Loss of job and income has intensified food insecurity, health services have been interrupted in parts or entirely around the world, and many policies have had negative mental and physical health consequences (WHO 2022). Global lockdowns, isolation, millions of deaths, and millions more affected, directly or indirectly, resulted in not only a substantial alteration in society and individual behaviour but also a shift in spatial perception (Bil et al. 2021). Despite the fact that highly concentrated mental health services, particularly in high- and middle-income societies, have become virtual, lockdowns, particularly in urbanised zones, have become a trigger in the deterioration of many people's mental health (Bil et al. 2021). People's mental health was deteriorating as in-person contact was lost and practically every element of life became virtual (Bil et al. 2021). Given that COVID-19 has had the most impact in urban areas, actions in cities and towns are likely to have the greatest impact. The COVID-19 infection appears to be influenced by urban morphology, with radial and grid cities having higher infection rates than linear cities (AbouKorin et al. 2021).

City centres, major urban arteries (high streets or main commercial avenues), and urban streetscapes are already being impacted by the pandemic (Florida et al. 2021). Lockdowns and the shift to delivery-based consumption have struck a tremendous blow to the already ailing urban retail sector, causing numerous enterprises to contract and fail (Florida et al. 2021). High streets may lose their commercial function, becoming instead amenity centres and window-shopping locations, luring shoppers to shop online (Florida et al. 2021). The pandemic is predicted to have a significant impact on how cities are maintained and governed in the future (Sharifi and Khavarian-Garmsir 2020). Actions taken in the coming years will determine whether post-COVID cities will be constructed and managed in a more sustainable manner (Sharifi and Khavarian-Garmsir 2020). Pandemics and other disruptions, such as wars, natural disasters, economic crises, and scientific advancements, have all occurred in the past (Fiorentino and Allmendinger 2021). While cities have survived, such calamities have dramatically affected the way we live and perceive them (Fiorentino and Allmendinger 2021). As a result, COVID-19 and the global pandemic may alter our future urban lifestyles (Fiorentino and Allmendinger 2021). The global health pandemic demonstrated how green space availability is a key determinant of environmental justice (Herman and Drozda 2021). Past research highlighted that low-income people and people of colour, who are the communities severely afflicted by COVID-19, have limited access to green spaces (Spotswood et al. 2021). Furthermore, cities are particularly vulnerable to climate change and emit a disproportionate amount of CO_2 (Wahba 2021). However, this also means they are critical to long-term climate stability and the areas where green investments will have the most benefit (Wahba 2021). New approaches to city planning and design are required to improve citizens' physical and mental well-being as well as lifestyle. A post-pandemic recovery will necessitate innovative solutions that combine existing plans with research-based ideas to handle the current difficulties (FEMA 2021). To fulfil social distance requirements while still aiding economic recovery, communities will need to rethink public spaces (FEMA 2021). They will also have to think about changing the way buildings are used and making zoning modifications to accommodate these changes (FEMA 2021). Pandemics have historically had catastrophic repercussions, but they have also given opportunities to create changes that result in communities that are healthier, more equal and with more green spaces (FEMA 2021; Klein 2020). In this chapter, we have examined the contribution of the landscape architecture discipline that could contribute to the resilience and sustainability of cities after the pandemic.

2 What is Landscape Architecture?

Although Gilbert Laing Meason created the term landscape architecture in his 1828 book "On the Landscape Architecture of the Great Painters in Italy"(Meason 1828), many theorists and scholars of landscape architecture have presented multiple interpretations of what landscape architecture is today (Murphy 2016). The American Landscape Architect Norman T. Newton in his book "Design on the Land: The Development of Landscape Architecture" defines landscape architecture as "the art of arranging land, together with the spaces and objects upon it, for safe, efficient, healthful, pleasant human use"(Newton 1971). Murphy defines Landscape Architecture as "the design discipline dedicated to understanding and shaping the landscape […] The primary role of landscape architecture is to organize the complexity of the landscape into comprehensible, productive, and beautiful places to improve the function, health, and experience of life"(Murphy 2016). In addition, landscape architecture has two main objectives: (1) to direct change in the landscape's shape in order to build and preserve useful, healthy, and engaging built and natural settings, and (2) to protect and improve the landscape's fundamental cultural, ecological, and experiential values (Murphy 2016). According to the American Society of Landscape Architects "landscape architecture involves the planning, design, management, and nurturing of the built and natural environments. With their unique skill set, landscape architects work to improve human and environmental health in all communities. They plan and design parks, campuses, streetscapes, trails, plazas, residences, and other projects that strengthen communities" (ASLA 2022).

Landscape architecture has awoken from the long slumber it fell into during the Modernist era and remained in for the majority of the twentieth century (Merkel 2007). In the last two decades, landscape architecture has undergone a paradigm shift, pushing designers to respond to the dynamic and temporal features of the landscape (Cantrell and Holzman 2015). Within landscape architecture, a generational movement has arisen that emphasises "distanced authorship" emphasising natural processes like succession, accretion, or passive remediation as landscape design agents (Cantrell and Holzman 2015).

This means that designers are exploring different ways of approaching their role in landscape architecture, specifically recognising the inherent dynamism and resilience of ecosystems. They aim to work with, rather than against, these natural processes. Professor Charles Waldheim identifies three main approaches (Waldheim 2006).

The first approach is rooted in critical architectural discourse and involves the use of automated methods to create highly sculpted surfaces in urban landscape projects. These projects reflect the continuation of the neo-avant-garde architectural movement (Waldheim 2006).

The second approach focuses on creating open-ended and indeterminate urban projects, distancing themselves from questions of authorship. These projects aim to be socially responsible, economically efficient, and adaptable to future urban forces. Examples include projects by architects such as Bernard Tschumi and Rem Koolhaas/ OMA, which emphasise the concept of open works and unplanned juxtapositions (Waldheim 2006).

The third approach explicitly incorporates ecological considerations. These projects use natural processes, landscape strategies, or ecological regimes as initial steps towards future urban development. They highlight the autonomy of ecological systems and their potential to shape urbanisation. An example is James Corner/ Field Operations' proposal for Lardner Point in Philadelphia, where contaminated soil is cleaned through phytoremediation, and the resulting urban voids inform future development (Waldheim 2006).

Landscape architects are creating new kinds of parks on brownfields, along waterfronts, even on rooftops and in rubbish dumps, thanks to the environmental movement, collaborations with architects and artists, and opportunities to regenerate abandoned industrial sites (Merkel 2007). Landscape architecture in the contemporary landscape project dimension focuses on sustainability by reducing resource waste and taking into account the effects of the climate emergency, which determines an evolution in spatial composition and formal outcomes. Nowadays, landscape architects address significant issues such as rapid urbanisation, social integration, rural–urban migration, offering vital and climate-resilient landscapes, planning adaptations to natural disasters and restoring degraded and polluted urban environments (Holden and Liversedge 2014). In particular, the landscape is at the centre of the great urban transformations. According to Michael Grove, Chair of Landscape Architecture, Civil Engineering, and Ecology, who sits on Sasaki's Board of Directors, landscape architecture should focus on four areas, each of which landscape architects have the unique talent to lead to be the first line of defence for preventing future pandemics (Grove 2020). These include: (1) promoting habitat conservation; (2) combating climate change; (3) promoting responsible urbanisation and reducing sprawl; and (4) promoting advanced agriculture (Grove 2020).

3 Landscape Architecture Promotes Healthy, Climate-Smart, and Resilient Communities

The purpose of resilient landscape architecture is to retrofit our communities such that they can recover faster from extreme shocks in the present and future (King and Green 2022).

Extreme shocks, such as flooding, severe storms, and climate change, pose clear dangers. Individuals, groups, regions, and countries are all affected. There is an opportunity to integrate resilient design concepts and practises with innovative design and landscape architecture to attenuate such shocks while promoting community investment and regeneration (Watson and Adams 2012). Landscape architects are assisting in the transition to a carbon-free society (Wright 2022). They plan and create dense, walkable communities to reduce transportation and sprawl-related pollution. Green roofs, water-efficient design, and the use of sustainable materials and construction processes are examples of how they make the built environment more energy and carbon-efficient (Wright 2022). They contribute to reducing atmospheric carbon dioxide by protecting and expanding carbon-sequestering habitats including forests, wetlands, and grasslands (Wright 2022). All of these measures help communities adapt to climate change and build resilience (Wright 2022).

Landscape architecture for climate change adaptation is being implemented in Copenhagen's St. Kjelds neighbourhood (klimakvarter 2016). A climate change adaptation strategy, complete with visions, strategies, and initiatives, was presented in 2013 (Kjaer 2015). The objective is to address the issues that Skt. Kjeld's neighbourhood faces by providing the neighbourhood with the identity that it has lacked due to climate change adaption. Furthermore, the objective is to turn rainwater into a positive element in the city and to manage it in a way that minimises damage while still providing new experiences in the neighbourhood (Kjaer 2015). Adaptation activities were guided by five objectives:

1. Separate the sewage system from 30% of the area.
2. Try to keep as much water in the upper catchment region as feasible.
3. Keep surface temperatures low.
4. Provide pathways that can lead away from the water in the event of severe rain.
5. Boosting biodiversity (Kjaer 2015).

When there are cloudbursts, the projects are designed to divert rainwater away from streets, homes, and basements, but they are also designed to create lush and attractive urban places for Copenhagen residents to enjoy when the weather is nice (klimakvarter 2016).

The first project that was developed is Tåsinge Plads, Copenhagen's first climate-friendly urban space, formally inaugurated in December 2014 (klimakvarter 2016). The area was built as part of "Sharing Copenhagen", a festival in which Copenhagen, as the European Green Capital, showcased the greatest solutions to future environmental and climate concerns (klimakvarter 2016) (Fig. 1).

The square is both a response to a demand for rainwater management and an investment in a better urban living (klimakvarter 2016). During rain events that occur once a year, the rain bed on the square will be filled up to 10, 30% during rain events that occur once every 25 years, and 40% during rain events that occur once every 100 years (Klimakvarter 2015). Extreme precipitation is also taken into consideration by the retention basin on Tåsinge Plads (Klimakvarter 2015). The retention basin (raingarden) will be completely filled in the event of cloudbursts that occur once every 500 years (Klimakvarter 2015). When the volume of water exceeds

Fig. 1 Tåsinge Plads, Copenhagen's first climate-friendly urban space (Image by Leif Jørgensen, Own work, CC BY-SA 4.0, https://commons.wikimedia.org/w/index.php?curid=52452785)

the capacity of the system, rainwater will flow from Tåsinge Plads to Tåsingegade's cloudburst street (Klimakvarter 2015).

In America, the transdisciplinary Blue Ribbon Panel on Climate Change and Resilience was convened in September 2017 by the American Society of Landscape Architects (ASLA 2018). A varied group of practitioners, experts, and stakeholder representatives served on the panel, each having expertise working at various scales in a variety of geographic and technological sectors (ASLA 2018). The panel was given two assignments: the first was to identify the most significant design and planning techniques for developing healthy, climate-smart, and resilient communities, and the second was to provide specific public policy suggestions to support those approaches (ASLA 2018). Their core principles, key planning and design strategies, and recommendations are summarised in the report "Smart Policies for a Changing Climate", which is a blueprint for ensuring a long-term, sustainable, and resilient future (ASLA 2018). The report identifies the following core principles as a foundation for resilient public policies.

1. Wherever possible, policies should be incentive-based.
2. Policies should encourage holistic planning and offer many advantages.
3. Policies should address issues of environmental justice as well as racial and socioeconomic equity.
4. Policies should be based on real community participation.

5. Policies should be evaluated against performance measures and checked for unintended consequences on a regular basis (ASLA 2018). The key design strategies are summarised in Table 1.

Table 1 Key design strategies adapted from ASLA 2018

Strategy	Where	Recommendations
Green infrastructure	New and existing urban and suburban developments	Green infrastructure should be incorporated into all new and existing urban and suburban developments
Reduce paved area	Streets	Reduce paved surfaces on the street and incorporate more trees and vegetation supported by healthy soils, such as bioswales and rain gardens
Green roofs	Urban rooftops	Encourage the use of green roofs for rainwater management and air purification
Porous pavement technologies	Built environment	To enhance natural hydrology, increase the use of porous pavement technology
Tree canopy	Built environment	Consider tree preservation and enhancement, as well as the increase of tree canopy
Stormwater	Built environment	Use cisterns to collect and reuse excess stormwater for irrigation and other purposes
Water	All development and land uses	Water conservation and reuse technology should be included into all development and land uses
Water sources	Built environment and beyond	Using the best research evidence in conjunction with design and planning practises, protect vital water sources, including aquifers
Plant selection	Built environment	Consider native or adapted plant species that are biohabitat-supporting and pollinator-friendly, and that are suitable for the site/region and changing climate conditions
Greenways and wildlife corridors	Regional	Greenways and wildlife corridors should be included in regional planning to enhance plant and animal migration and relocation

4 Green Recovery and Landscape Architecture

Because the COVID-19 outbreak occurred at a time when environmental sustainability and climate change were at an all-time high, many policy organisations and academics across Europe, North America, and other nations, promoted a narrative focused on a "green recovery" that would boost the economy and accelerate low-carbon transitions (European Commission 2020; Geels et al. 2022). Landscape architecture could play a key role in a "green recovery" after the pandemic as well as in achieving the Sustainable Development Goals by 2030. According to a policy paper published by the Landscape Institute (LI) in the United Kingdom, COVID-19 pandemic related to economic and social recovery must be green (Landscape Institute 2020b). British landscape architects argue that the phrase "green recovery" cannot only be a slogan, but the UK needs to focus on five key elements in order to establish a really sustainable recovery (Landscape Institute 2020b):

(1) Design new infrastructure and housing with a natural capital mindset.
(2) Invest in the maintenance and renovation of existing structures.
(3) Develop more stringent and equitable green space criteria.
(4) Start investing into natural climate change solutions.
(5) Achieve a paradigm shift in green skills, data, and digital (Landscape Institute 2020b).

The LI and The Parks Alliance (TPA) recognise that parks and green spaces play an important role in delivering a greener recovery (Landscape Institute 2020a). Parks and green areas received a lot of attention during the COVID-19 lockdown, and they were rightfully lauded as important assets for people's physical and mental well-being (Landscape Institute 2020a). They became an important part of the UK's pandemic response (Landscape Institute 2020a). People have turned to local parks like never before during the pandemic for fresh air, exercise, meditation, and a sense of tranquillity (Trust for Public Land 2020). Parks and public spaces, in particular, are seeing unprecedented high levels of use, with municipal officials reporting significant increases in visitors (Kleinschroth and Kowarik 2020; Trust for Public Land 2020; Venter et al. 2020). For example, residents of New York City are using urban green spaces far more frequently now than they were during the lockdown or before the outbreak (Pipitone and Jović 2021). However, a study conducted in Krakow, Poland, showed that the number of persons who visited green spaces during the pandemic fell to 78.9% of the population, down 13.1% from pre-pandemic levels (Noszczyk et al. 2022). In Saudi Arabia, Addas and Maghrabi (2022) published similar findings, revealing that more than 65% of survey respondents never visited parks during the pandemic (Addas and Maghrabi 2022). Urban green spaces and parks provide essential ecosystem services (Russo and Cirella 2020, 2021) that help people cope with COVID-19-related stress (Geng et al. 2021). Several studies reported that COVID-19 limitations and closures of parks and green spaces have been proven to cause serious physical and mental health problems in people, particularly those living in metropolitan areas (Geng et al. 2021; Slater et al. 2020).

In light of the numerous health benefits, many outdoor spaces reopened with physical distance rules and other measures after research found that the risk of COVID-19 transmission was much lower in urban green spaces, giving a safe and much-needed refuge for many individuals living in lockdown (Larson et al. 2022; Slater et al. 2020). Green space preservation and development has been highlighted in research as a strategy to sustain mental health, particularly during psychologically stressful circumstances (Vos et al. 2022). According to research, urban green spaces helped to mitigate the decline in physical activity that occurred throughout the pandemic (Yang et al. 2021). During the COVID-19 outbreak in China, Chinese scientists found that parkland provision was favourably connected with mental well-being (Yao et al. 2022). People in Spain who were under rigorous lockdown reported that nature helped them cope with the measures and that people having access to outdoor spaces and blue-green elements in their views had more pleasant emotions (Pouso et al. 2021). A study conducted in Israel during the lockdown, reported that all respondents' exposure to nature was limited, and some indicators of personal and neighbourhood well-being were poorer for those who lived in grey neighbourhoods (Colléony et al. 2022). A study conducted in the United Kingdom has revealed that an overwhelming 90% of participants acknowledged the positive impact of green spaces on their quality of life during the pandemic (Crossley and Russo 2022). Therefore, extensive scientific literature strongly supports the idea that having access to green spaces brings numerous advantages, enhancing the quality of life, promoting human health and well-being, and concurrently reducing healthcare expenses (WHO 2016, 2021).

According to some recent estimates in England, £2.1 billion in annual health costs could be saved if everyone in England had good access to greenspace, due to increased physical activity in those spaces; the annual net benefit to society of Birmingham's parks and greenspace is nearly GB£600 million, including GB£192 million in health benefits (Public Health England 2020). In Sheffield, every GB£1 invested in park maintenance saves GB£34 in medical costs, with local residents benefiting the most (Public Health England 2020). A green recovery could be also an opportunity to enhance and increase green spaces and rewild cities, for example already in the UK one-quarter of English councils have plans to rewild their urban spaces (Yeo 2021). Doncaster Borough Council is expected to include a review of its green space management regimes to allow them to become "wilder". The council has set aside GB£300,000 for "capital resources for the acquisition of equipment and other materials" (Yeo 2021). A notable example is the "Green Minds Plymouth" project, led by Plymouth City Council, which is testing diverse approaches to rewilding, with a focus on incorporating communities and social entrepreneurs (Yeo 2021). Current initiatives include beaver reintroductions and urban rewilding (Yeo 2021) (Fig. 2).

Rewilding also represents an opportunity to regenerate urban brownfield (Masood and Russo 2023; Russo and Cirella 2017) or abandoned spaces such as shopping centres; for example, as part of a post-pandemic urban rewilding project, a derelict 1970s era shopping centre in Nottingham might be turned into wetlands, pocket forests, and a wildflower meadow (Weston 2020). Many local governments are exploring innovative strategies to expand and improve their greenspaces, including

Fig. 2 Urban rewilding in Plymouth: Since 2014, Plymouth City Council has been involved in urban rewilding initiatives through the "Grow Wild" project with Kew Gardens, creating wildflower meadows. Today, the City Council continues its efforts with the "Green Minds" project, exploring various rewilding strategies and fostering community and social enterprise involvement (Yeo 2021) (Image by Alessio Russo)

using legislative mandates, building partnerships, collaborating with local communities, and utilising development agreements (Association for Public Service Excellence 2020). Recent biodiversity net gain agreements have mandated that future developments leave biodiversity in a better state than it was previously (Association for Public Service Excellence 2020). Developers are encouraged to propose projects that give an overall increase in natural habitat and ecological features to urban greenspaces when employing biodiversity net gain principles (Association for Public Service Excellence 2020). The post-pandemic phase presents an opportunity to establish national standards for the amount, quality, and accessibility of green infrastructure (Russo and Cirella 2018; Cirella 2020, 2022), such as parks and green spaces, as well as equity measures to ensure that the needs of local communities are addressed (Landscape Institute 2020a).

5 Conclusion

Following Covid, landscape architecture has come to be seen as an essential discipline for facilitating social contact as well as infection prevention (Quattro 2021). Responsible landscape design can improve social interaction and foster a sense of community (Quattro 2021). The pandemic has served as a catalyst for increasing public awareness about the significance of preserving and restoring our natural environments (Quattro 2021). In comparison to other related subjects such as urban planning and environmental planning, the concept of urban resilience has received little attention in the field of landscape architecture (Han et al. 2021). This chapter aimed to fill a gap in the literature by demonstrating how landscape architecture can help cities recover after a pandemic and prepare for future stress. The number of resilient urban regeneration projects is projected to rise. Landscape architecture is strongly linked to increasing people's health and well-being, regenerating cities, and improving the urban living environment.

As a response to the pandemic, people's values for greenspaces may shift, increasing pressure on cities to improve their urban settings through social and political initiatives (Kleinschroth and Kowarik 2020). Building healthy communities necessitates investment in public urban landscapes and urban landscape systems (van Ryzin 2021). Urban greenspaces can not only provide a stress-relieving environment during times of societal stress, such as a pandemic, but they can also assist individuals in physically and mentally preparing for such events (Kleinschroth and Kowarik 2020). By offering conditions for community interaction and social well-being, public spaces will play a critical role in reviving our cities after the COVID-19 pandemic (Innocent and Stevens 2021). Furthermore, through the implementation of edible green infrastructure, landscape architecture could play a strategic role in addressing future environmental shocks and food insecurity (Russo and Cirella 2019; Sardeshpande et al. 2021). According to Phoebe Lickwar, an Associate Professor in Landscape Architecture, the School of Architecture, University of Texas at Austin, urban agriculture should be integrated into the urban fabric in post-pandemic public spaces (van Ryzin 2021). Moreover, "we need to get serious about developing an urban agriculture that is equitable and capable of provisioning the city, not on its own, but in conjunction with other forms of rural agriculture"(van Ryzin 2021). Future research should focus on adaptive landscape design strategies at various scales, which can help to mitigate the effects of future shocks.

References

AbouKorin SAA, Han H, Mahran MGN (2021) Role of urban planning characteristics in forming pandemic resilient cities – case study of Covid-19 impacts on European cities within England, Germany and Italy. Cities 118:103324. https://doi.org/10.1016/j.cities.2021.103324

Addas A, Maghrabi A (2022) How did the COVID-19 pandemic impact urban green spaces? A multi-scale assessment of Jeddah megacity (Saudi Arabia). Urban Urban Green 69:127493. https://doi.org/10.1016/j.ufug.2022.127493

ASLA (2018) Smart policies for a changing climate: the report and recommendations of the ASLA blue ribbon panel on climate change and resilience smart policies for a changing climate. Washington DC

ASLA (2022) What is landscape architecture? https://www.asla.org/aboutlandscapearchitecture.aspx. Accessed 6 Feb 2022

Association for Public Service Excellence (2020) Post pandemic: green urban spaces - a brief guide to increasing the opportunities to green urban spaces and understand and value the benefits this will bring. Stretford

Bil JS, Buława B, Świerzawski J (2021) Mental health and the city in the post-COVID-19 era. Sustainability 13:7533. https://doi.org/10.3390/su13147533

Cantrell BE, Holzman J (2015) Responsive landscapes: strategies for responsive technologies in landscape architecture, 1st edn. Routledge, New York

Cirella GT (2020) Sustainable human-nature relations: environmental scholarship, economic evaluation, urban strategies. Springer, Singapore

Cirella GT (2022) Human settlements: urbanization, smart sector development, and future outlook. Springer, Singapore

Colléony A, Clayton S, Shwartz A (2022) Impacts of nature deprivations during the COVID-19 pandemic: a pre-post comparison. Biol Conserv 268:109520. https://doi.org/10.1016/j.biocon.2022.109520

Crossley AJ, Russo A (2022) Has the pandemic altered public perception of how local green spaces affect quality of life in the United Kingdom? Sustainability 14:7946. https://doi.org/10.3390/su14137946

European Commission (2020) Boosting the EU's green recovery: Commission invests €1 billion in innovative clean technology projects. https://ec.europa.eu/commission/presscorner/detail/en/IP_20_1250. Accessed 13 Feb 2022

FEMA (2021) Local government solutions guide for COVID-19 and Beyond: adaptive design

Fiorentino S, Allmendinger P (2021) Will Covid-19 change the way we live in our cities? https://www.alexandrinepress.co.uk/Post-Covid-Planning. Accessed 5 Feb 2022

Florida R, Rodríguez-Pose A, Storper M (2021) Cities in a post-COVID world. Urban Stud 004209802110180. https://doi.org/10.1177/00420980211018072

Geels FW, Pereira GI, Pinkse J (2022) Moving beyond opportunity narratives in COVID-19 green recoveries: a comparative analysis of public investment plans in France, Germany, and the United Kingdom. Energy Res Soc Sci 84:102368. https://doi.org/10.1016/j.erss.2021.102368

Geng D (Christina), Innes J, Wu W, Wang G (2021) Impacts of COVID-19 pandemic on urban park visitation: a global analysis. J For Res (Harbin) 32:553–567. https://doi.org/10.1007/s11676-020-01249-w

Grove M (2020) Essential workers: 4 ways landscape architects can be the first line of defense for preventing future pandemics. WLA. https://worldlandscapearchitect.com/essential-workers-4-ways-landscape-architects-can-be-the-first-line-of-defense-for-preventing-future-pandemics/. Accessed 3 Apr 2022

Han S, Sim J, Kwon Y (2021) Recognition changes of the concept of urban resilience: moderating effects of COVID-19 pandemic. Land (basel) 10:1099. https://doi.org/10.3390/land10101099

Herman K, Drozda Ł (2021) Green infrastructure in the time of social distancing: urban policy and the tactical pandemic urbanism. Sustainability 13:1632. https://doi.org/10.3390/su13041632

Holden R, Liversedge J (2014) Landscape architecture: an introduction. Laurence King Publishing Ltd., London

Innocent T, Stevens Q (2021) Urban play as catalyst for social wellbeing post-pandemic. Front Comput Sci 3. https://doi.org/10.3389/fcomp.2021.634145

King A, Green J (2022) Professional practice: resilient design. https://www.asla.org/resilientdesign.aspx#:~:text=Resilientlandscapeplanninganddesignoffersawayforwardforcommunities&text=Thegoalofresilientlandscape,nowandinthefuture. Accessed 17 Feb 2022

Kjaer HL (2015) Climate change adaptation in marginalized neighbourhoods. https://projekter.aau.dk/projekter/files/213985464/artikel.pdf. Accessed 30 Jun 2023

Klein C (2020) How pandemics spurred cities to make more green space for people. https://www.history.com/news/cholera-pandemic-new-york-city-london-paris-green-space. Accessed 20 Oct 2020

Kleinschroth F, Kowarik I (2020) COVID-19 crisis demonstrates the urgent need for urban greenspaces. Front Ecol Environ 18:318–319. https://doi.org/10.1002/fee.2230

Klimakvarter (2015) Tåsinge plads: a green oasis in the climate resilient neighbourhood. A place where rainwater sets the scene for play and social interaction.

klimakvarter (2016) Copenhagen's First Climate resilient neighbourhood

Landscape Institute (2020a) Greener recovery-delivering a sustainable recovery from COVID-19. London

Landscape Institute (2020b) A green recovery for parks and green spaces: LI policy paper (supplement). London

Larson LR, Mullenbach LE, Browning MHEM et al (2022) Greenspace and park use associated with less emotional distress among college students in the United States during the COVID-19 pandemic. Environ Res 204:112367. https://doi.org/10.1016/j.envres.2021.112367
Masood N, Russo A (2023) Community perception of brownfield regeneration through urban rewilding. Sustainability 15:3842. https://doi.org/10.3390/su15043842
Meason GL (1828) On the landscape architecture of the great painters of Italy. Printed at C. Hullmandel's Lithographic Establishment, 51, Gt. Marlborough Street
Merkel J (2007) Urban American landscape. Archit Des 77:36–47. https://doi.org/10.1002/ad.422
Murphy MD (2016) Landscape architecture theory: an ecological approach, 1st edn. Island Press/ Center for Resource Economics, Washington, DC
Newton NT (1971) Design on the land: the development of landscape architecture. Belknap Press, USA
Noszczyk T, Gorzelany J, Kukulska-Kozieł A, Hernik J (2022) The impact of the COVID-19 pandemic on the importance of urban green spaces to the public. Land Use Policy 113:105925. https://doi.org/10.1016/j.landusepol.2021.105925
Pipitone JM, Jović S (2021) Urban green equity and COVID-19: effects on park use and sense of belonging in New York City. Urban Urban Green 65:127338. https://doi.org/10.1016/j.ufug.2021.127338
Pouso S, Borja Á, Fleming LE et al (2021) Contact with blue-green spaces during the COVID-19 pandemic lockdown beneficial for mental health. Sci Total Environ 756:143984. https://doi.org/10.1016/j.scitotenv.2020.143984
Public Health England (2020) Improving access to greenspace. A new review for 2020. London
Quattro (2021) How is landscape architecture responding to the pandemic? https://www.quattrodesign.co.uk/how-is-landscape-architecture-responding-to-the-pandemic/. Accessed 27 Feb 2022
Russo A, Cirella GT (2017) Biophilic cities: planning for sustainable and smart urban environments. In: Aijaz R (ed) Smart cities movement in BRICS. Observer Research Foundation and Global Policy Journal, New Delhi, pp 153–159.
Russo A, Cirella GT (2018) Modern compact cities: how much greenery do we need? Int J Environ Res Public Health 15:2180. https://doi.org/10.3390/ijerph15102180
Russo A, Cirella GT (2019) Edible urbanism 5.0. Palgrave Commun 5:163. https://doi.org/10.1057/s41599-019-0377-8
Russo A, Cirella GT (2020) Urban sustainability: integrating ecology in city design and planning. In: Cirella GT (ed) Sustainable human-nature relations: environmental scholarship, economic evaluation, urban strategies. Springer, Singapore, pp 187–204
Russo A, Cirella GT (2021) Urban ecosystem services: current knowledge, gaps, and future research. Land (basel) 10:811. https://doi.org/10.3390/land10080811
Sardeshpande M, Rupprecht C, Russo A (2021) Edible urban commons for resilient neighbourhoods in light of the pandemic. Cities 109:103031. https://doi.org/10.1016/j.cities.2020.103031
Sharifi A, Khavarian-Garmsir AR (2020) The COVID-19 pandemic: Impacts on cities and major lessons for urban planning, design, and management. Sci Total Environ 749:142391. https://doi.org/10.1016/j.scitotenv.2020.142391
Slater SJ, Christiana RW, Gustat J (2020) Recommendations for keeping parks and green space accessible for mental and physical health during COVID-19 and other pandemics. Prev Chronic Dis 17. https://doi.org/10.5888/PCD17.200204
Spotswood EN, Benjamin M, Stoneburner L et al (2021) Nature inequity and higher COVID-19 case rates in less-green neighbourhoods in the United States. Nat Sustain 4:1092–1098. https://doi.org/10.1038/s41893-021-00781-9
Trust for Public Land (2020) Parks and the pandemic
van Ryzin JC (2021) Edible cities: landscape architect Phoebe Lickwar on post-pandemic public spaces and urban agriculture. Sightlines
Venter ZS, Barton DN, Gundersen V et al (2020) Urban nature in a time of crisis: recreational use of green space increases during the COVID-19 outbreak in Oslo, Norway. Environ Res Lett 15:104075. https://doi.org/10.1088/1748-9326/abb396

Vos S, Bijnens EM, Renaers E et al (2022) Residential green space is associated with a buffering effect on stress responses during the COVID-19 pandemic in mothers of young children, a prospective study. Environ Res 208:112603. https://doi.org/10.1016/j.envres.2021.112603
Wahba S (2021) Future of cities will shape post-COVID-19 world. https://www.worldbank.org/en/news/feature/2021/03/02/future-of-cities-will-shape-post-covid-19-world. Accessed 5 Feb 2022
Waldheim C (2006) Strategies of indeterminacy in recent landscape practice. Public
Watson D, Adams M (2012) Design for flooding: architecture, landscape, and urban design for resilience to flooding and climate change
Weston P (2020) Going wild? A radical green plan for Nottingham's unloved shopping centre. The Guardian
WHO (2016) Urban green spaces and health: a review of the evidence. Copenhagen
WHO (2021) Green and blue spaces and mental health: new evidence and perspectives for action. WHO Regional Office for Europe, Copenhagen
WHO (2022) COVID-19 and urban health. https://www.who.int/teams/social-determinants-of-health/urban-health/covid-19. Accessed 5 Feb 2022
Wright A (2022) Climate change mitigation and landscape architecture. https://www.asla.org/climatemitigation.aspx. Accessed 27 Feb 2022
Yang Y, Lu Y, Yang L et al (2021) Urban greenery cushions the decrease in leisure-time physical activity during the COVID-19 pandemic: a natural experimental study. Urban Urban Green 62:127136. https://doi.org/10.1016/j.ufug.2021.127136
Yao Y, Lu Y, Guan Q, Wang R (2022) Can parkland mitigate mental health burden imposed by the COVID-19? A national study in China. Urban Urban Green 67:127451. https://doi.org/10.1016/j.ufug.2021.127451
Yeo S (2021) One quarter of English councils have plans to rewild. Does yours? Inkcap J

Transformation of Public Spaces Through Emergency Urbanism: Lessons from Toronto, Recife, and New York

Maria de Lourdes Carneiro da Cunha Nóbrega, Isabella Leite Trindade, and Ana Luisa Rolim

Abstract This chapter presents a comparative analysis of urban interventions involving the use of public spaces that took place during the pandemic from May 2020 to July 2021 in Toronto (Canada), Recife (Brazil), and New York (USA). The identification and analysis of these transformations can contribute to the constitution of different regulations and urban plans, which go beyond the context of pandemics, supporting sustainable urbanism practices essential to the (re)balance of the ecosystem currently threatened by the appearance and proliferation of the coronavirus disease (COVID-19). The interventions observed in each city are based on the following conceptual foundations: principles of sustainable urbanism; research on the use of open spaces versus the detriment of closed spaces during the pandemic; recommendations and data from the World Health Organization; and definitions on tactical urbanism, and emergency urbanism. The analysis presented here focuses on practices of tactile urbanism (or "placemaking" or emergency urbanism), the understanding of the need for paradigm shifts for the use of public spaces in the face of the COVID-19 pandemic, and the urgency to present urban actions in response to the new restrictions imposed. For this purpose, the following ways of using the city were observed: circulation, permanence, and commerce. It is argued that these three levels of action, acting together, generate conditions for the use of public space that strengthen the practice of sustainable urbanism in contemporary times.

Keywords COVID-19 pandemic · Emergency urbanism · Public spaces · Tactile urbanism · Urban sustainability

M. de Lourdes Carneiro da Cunha Nóbrega (✉) · A. L. Rolim
Department of Architecture and Urbanism, Catholic University of Pernambuco, Boa Vista, Recife 50050-900, Brazil
e-mail: lourdinha_@hotmail.com

A. L. Rolim
e-mail: ana.rolim@unicap.br

I. L. Trindade
School of Interior Design, Ryerson University, Toronto, ON M5B 2K3, Canada

G. T. Cirella and B. Dahiya (eds.), *City Responses to Disruptions in 2020*, Advances in 21st Century Human Settlements, https://doi.org/10.1007/978-981-99-7988-2_4

1 Introduction

The coronavirus disease (COVID-19) caused by the new coronavirus sars-CoV-2, began to spread in mid-December 2019 from Wuhan, China. First, to cities within China itself, then from January 2020 to more than 200 other countries, being declared a worldwide pandemic in March 2020. Due to the pandemic, safety protocols were adopted to contain the spread of the virus among the population, with changes in habits, such as: wearing masks, washing hands, frequent use of hand sanitizer, cleaning and disinfecting high-touch surfaces daily, and keeping social distance from others outside of our home. The acronym WFH (work from home) changed to WFE (work from everywhere), as part of the population started to work remotely, from anywhere. Events with large gatherings were canceled, businesses and services considered non-essential were temporarily closed, and educational institutions started to teach remotely as well. At the same time, cities have implemented safety regulations and protocols for the use of public space, with varying restrictions, and total or partial closure of activities in these places. Thus, at the beginning of the pandemic, we started to see a shift in how people use public space. Among many changes in a short period, a change in daily life was noticeable, leaving cities empty due to the measures adopted to contain the virus (Kahraman et al. 2022).

Starting from this initial context, which required safety protocols, recommendations, and hygiene measures, in addition to normative adjustments for the use of public space, the objective of this chapter is to present and analyze urban interventions that took place in the cities of Toronto (Canada), Recife (Brazil) and New York (USA),[1] involving the use of public space during the COVID-19 pandemic. The identification and analysis of these transformations can, for example, contribute to the constitution of different regulations and urban plans, which go beyond the context of pandemics, supporting sustainable urbanism practices essential to the equilibrium of the ecosystem, which is currently threatened by the appearance and proliferation of the virus. The interventions observed in each city between May 2020 and July 2021 are theoretically based on the following references: principles of sustainable urbanism by Farr (2013); research on the use of open spaces versus the detriment of closed spaces during the COVID-19 pandemic (Bazant and Bush 2021); recommendations and data from the World Health Organization—WHO (2021); definitions on tactical urbanism, emergency urbanism and practices related to informational space (Bambozzi 2019) and, specifically for New York City, the "NYC's Open Restaurant

[1] Taking into account these different socioeconomic realities, the number of vaccinated people is an indicator of the possibility of greater or lesser use of public space. The percentage of vaccinated people for the three cities at the end of the period of on-site observations of this research (June 30, 2020) is presented for comparative purposes: Toronto: 81.61% with the first dose and 46.86% with the second dose (source: https://www.toronto.ca/home/covid-19/covid-19-latest-city-of-toronto-news/covid-19-pandemic-data/covid-19-weekday-status-of-cases-data/); Recife: 39.2% with the first dose and 14.5% with the second dose (source: https://conectarecife.recife.pe.gov.br/vacinometro/); New York: 55.72% with the first dose and 50.74% with the second dose (source: https://www1.nyc.gov/site/doh/covid/covid-19-data-totals.page). It is also understood that the greater the number of people vaccinated, the greater the possibility of agglomerations in each city.

Program" regulation on the use of public streets for restaurants (NYC DOT 2021a, b, c, d).

The analysis presented here focuses on the practices of tactile urbanism (or "place-making"), emergency urbanism and the understanding of the need for paradigm shifts for the use of public space in the face of a pandemic. For this purpose, we argue that three levels of action—circulation, permanence, and commerce—generate conditions for the use of public space that strengthen the practice of sustainable urbanism in contemporary times.

2 Urban Interventions in Emergency Situations

The basic principles of sustainability (Nações Unidas Brazil 2020) guide urban planning practices at the beginning of the twenty-first century. Urban planning that was previously directed towards the use of the individual automobile, and zoning, which fragment and spread the urban fabric, gave way to new forms of urban design and planning. According to Farr (2013, pp. 30–40), new urban practices must pursue urban sustainability through actions that aim to achieve the following five principles: (1) "compactness", which goes against urban sprawl, in search of population and constructive densities typically balanced in cities; (2) the "completeness", which proposes a mix of uses, always at the hand of the urban user; (3) "connectivity", which allows access to the uses proposed by completeness in urban areas, preferably on foot and/or through the use of collective and/or non-motorized transport, to the detriment of individual transport (Cirella et al. 2019; Russo and Cirella 2020; Ravindra et al. 2022); (4) "biophilia", which proposes the reintegration of urban inhabitants into nature, and the preservation of the natural environment; and (5) permeating all these principles, is "walkability", a measure of how friendly an area is for walking.

It is, therefore, through the emergence of the global pandemic that a paradox arises regarding the use of urban public space: on the one hand, the practice of sustainable urbanism requires basic urban actions that favor greater density, rapid connectivity between places, walkability, and the use of public transport, to the detriment of the use of individual transport. On the other hand, the contingencies for the use of these spaces and collective forms of locomotion, so as not to spread the virus, came to define them as areas susceptible to crowding of people and consequent risks to public health. How, then, could citizens live in urban centers safely while avoiding contagion? At the first moment of the pandemic, the general practice was for public authorities to adopt social distancing measures, lockdown, and/or "Stay at Home Orders", to prevent the spread of the virus, and only authorize the operation of essential services (supermarkets, pharmacies, and the like).

Thus, by adopting these measures, the principles of completeness and connectivity of urban space were affected through the emptying of streets and other spaces for public use. The lack of movement of people during the pandemic directly influenced the practices of retail trade and service activities—resulting in economic loss for these activities, which depend on the permanent flow of people for their economic sustainability. This paradox extends to the use of public transport, an essential means of transportation to guarantee the principle of urban connectivity, as well as walkability, since its users are essentially pedestrians. Closed vehicles (such as buses and subway trains), were avoided, as shown by the data from the Global Report on Public Transport[2] (Moovit 2021). This report shows that 36% of passengers reduced their use of public transport at the start of the pandemic; 10% migrated to other means of transport; and 4% stopped using it altogether.

Bazant and Bush (2021) show that increased virus transmission is associated with internal areas, and closed environments have the greatest risk of spreading the virus. They note that 80% of cases occurred inside these environments in events that are understood as "superscattering", where relatively small micron-scale aerosol droplets of the virus play a dominant role in the spread of COVID-19, citing, for example, the contamination of 23 out of 68 passengers on a 2-h bus trip in Ningbo, China (Bazant and Bush 2021). Also using a Chinese study that analyzed 7,324 COVID-19 cases, Qian et al. (2020) demonstrate that closed environments show the greatest risk of spreading the virus, noting that 80% of cases occurred inside homes and 34% involved public transport with only a single transmission found outdoors. Based on these studies, the use of open space is understood as an alternative to the practice of urban sustainability in response to the urban paradox created by the pandemic. To this end, urban centers started to host tactical urbanism (or placemaking) actions, or guerrilla urbanism (Malagi and Mehta 2020) or emergency urbanism, as an alternative to actions aimed at promoting principles of sustainability in urban spaces, thus minimizing economic and social losses during the pandemic.

Tactical urbanism consists of simple and low-cost actions (such as painting the street, adding street furniture, creating parklets, etc.), which require community participation and essentially "covers planning, design, management and the programming of public spaces" in order to facilitate "the creation of activities and connections (cultural, economic, social, environmental) that define a space and support its evolution" (Heemann; Santiago s/d, p. 10). As an emergency alternative plan for this pandemic period, empty streets gave rise to actions that would provide quick and low-cost alternatives to minimize the paradoxical issues that affect the use of urban space arising from the pandemic, where people could return to the streets, but, with safety and social distance. Street painting, a strategy widely used in the practice of tactical urbanism, contributed to the redefinition of these spaces, and also amplified the capacity for city streets and other spaces to serve as a vehicle to support information sharing given that streets now accommodate more space for pedestrians and cyclists, as demonstrated by the field of Sustainable Urbanism (Farr 2013).

In these cases, a street becomes an informational space with messages on its floor and/or projections on facades, which are photographed and/or filmed and thus transmitted and echoed via social networks and various media. For example, in June

---

[2] Ten Brazilian metropolitan regions are part of the report: Belo Horizonte, Brasília, Campinas, Curitiba, Fortaleza, Porto Alegre, Recife, Rio de Janeiro, Salvador, and São Paulo. The survey was answered by 13,000 users in these cities in November 2020, indicating best practices in public transport and micromobility (shared bicycles and scooters, among others). All data are anonymous (Moovit 2021).

Fig. 1 (*Left*) Image cropped from the video posted by Washington Mayor Muriel Bowser, on June 5, 2020. *Source* https://twitter.com/murielbowser/status/1268916115809488896, accessed in July 2021; (*right*) image retrieved from the Facebook page "Mídia NINJA" showing the street painting (in execution) of the movement "*Vidas Pretas Importam*" on November 21, 2020. *Source* https://www.facebook.com/MidiaNINJA/posts/2058170327674435, accessed in August 2021

2020, the mayor of Washington Muriel Bowser shared on Twitter that she painted the street with the words "Black Lives Matter", in support of the "Black Lives Matter" movement, with worldwide repercussions, including in Brazil, as can be seen on *Avenida Paulista* (SP) in the same year. Streets are thus given a new meaning, by appropriating and serving as a screen for transmitting messages as well as through photographs and films diffused through different media networks (Fig. 1).

Urbanists such as Speck (2016) and Ghel (2015) also reinforce these practices by proposing in their studies and projects that walking is an alternative *sine qua non* for the practice of contemporary urbanism, thus promoting the city as a meeting place for people. In this sense, urban public space becomes a stage for projecting actions that aim to achieve the principles of urban sustainability. Thus, street space, for example, previously interpreted through design practices aimed only at circulation, "as a two-dimensional surface where cars pass", thus becomes understood as "a basic unit of urban space through which people experience the city" that "adapt over time to favor environmental sustainability, public health, economic activity and cultural importance" (SENAC 2018, p. 4).

In the context of these practices, the urban environment characterized by Santos (1994, p. 24) as "technical-scientific-informational" began to expand its capacity for information through the digital world and social media. Spaces and places started to be characterized and re-signified as informational places (e.g., political places, places of social speech, and places of media organization), as well as the formation of new urban layers, based on digital information. The definition of informational space migrates between fields. To start, it is a geographical concept, but it has also been applied to architecture, art, and information. It is a web that unites visible and invisible fields and relationships that connect urban spaces, the specifics of place and context, telematic networks, the internet and radio, television, telephone, and other signal-inducing waves. These are signs that can be carefully measured or just assumed, effectively perceived through their physical impact or assumption, in more scientific or imagined ways, fanciful as they may be (Bambozzi 2019, p. 13).[3]

Therefore, when the actions and practices that constitute emergency urbanism (during a pandemic) are inserted into the practice of sustainable urbanism, these actions not only transform urban spaces, physically and functionally, but result in urban interventions adopted for the use of public space, as in the cities of Toronto (Canada), Recife (Brazil), and New York (USA). Advanced planning governs both intervention strategies whether through the implementation of tactical urbanism actions in search of a walkable city (as in the case of Recife) or the implementation of emergency urbanism (as in the case of Toronto and New York City). But it is noteworthy that planning, designing, and managing the use of public space in ways that make it possible to circulate directly contribute to sustainable urbanism practices, even during a global pandemic, as the economy and mental health of citizens are also taken into account. As life begins to return to normal or the "new normal", these projects and initiatives present an opportunity to assess our capacity for change and adaptation in the face of crises and produce creative alternatives by transforming space. Furthermore, they provide an opportunity to think about where we want to live, how we want to live, and how to overcome the current shortcomings of the city.

3 The Pandemic Context, the Transformation of Public Spaces: Toronto, Recife, and New York

3.1 Toronto

In Toronto, the lockdown and "stay at home order" was imposed on March 13, 2020, together with a set of initiatives that were created in response to the pandemic: ActiveTO, CaféTO and CurbTO programs. In addition to these programs, in a great display of community collaboration, five Toronto-based organizations—Cycle Toronto, Walk Toronto, The Bentway, Park People, and Spacing—have come together to advocate for safe and viable ways for Torontonians to navigate their city on foot and by active transportation. The initiative is centered around an emerging

[3] Original Portuguese quote: "A definição de espaço informacional migra entre campos. É a princípio um conceito geográfico, mas que vem sendo aplicado nas crescentes formas de fricção entre arquitetura, arte e informação. É uma trama que une campos visíveis e invisíveis e relações que conectam espaços urbanos, as especificidades do lugar e do contexto, as redes telemáticas, a internet e as ondas de rádio, televisão, telefonia e outras formas indutoras de sinais. São sinais que podem ser minuciosamente mensurados ou apenas presumidos, percebidos efetivamente através de seu impacto físico ou suposição, por formas mais científicas ou imaginadas, fantasiosas que sejam" (Bambozzi 2019, p. 13). Translated by the authors.

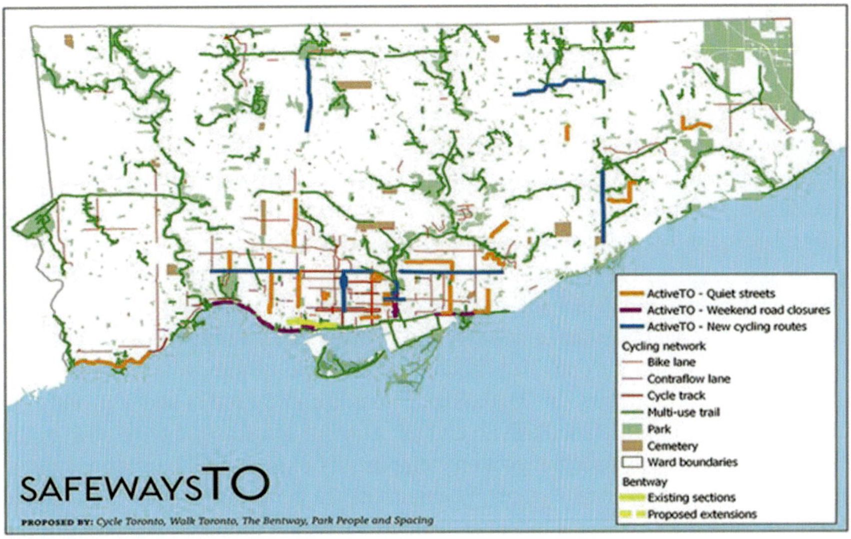

Fig. 2 SafewaysTO Map by Cycle Toronto, Walk Toronto, The Bentway, Park People, and Spacing. *Source* https://tinyurl.com/ydcayec2. Accessed in June 2020

network of bike lanes, multi-use trails, parks and ActiveTO closures for recreational and commuting purposes. These "safeways" are both a critical response to the pandemic and hold the key to a safe and sustainable post-COVID-19 future (Fig. 2).

These initiatives emerged from a consensus among specialists that, if, on the one hand, people were advised to isolate themselves at home (stay-at-home order), on the other hand, it was also important to guarantee everyone's well-being and mental health, by creating mechanisms for people to be able to leave the house, remain active and socially connected, in a safe way, even in the pandemic context (Greenberg 2021).

One of the first actions taken to combat the spread of COVID-19 by the City of Toronto was the creation of the ActiveTO program, which focuses on making sure that people have space to get around while respecting physical distancing. Initiatives include closing major roads for active transportation, expanding the cycling network, and implementing quiet streets by temporarily closing them. The City of Toronto understood that it would be difficult to comply with physical distancing guidelines (2 m distance between people) in residential areas where there are narrow sidewalks. Thus, to minimize the problem, Quiet Streets designated 65 km of streets in 31 locations as "shared space" to enable people to maintain physical distancing while walking, running, using mobility devices, and biking. On these streets, signs and temporary barricades indicated that the circulation of cars was limited, by encouraging slow, local vehicle access only, and creating "quiet streets". This would make it possible for people to walk not only on the sidewalk but on the road itself, in order to keep social distancing to a minimum.

In addition to these "quiet streets", major road closures were closed on weekends and holidays. These short-term closures also helped to provide more space for people to be active, enabling physical distancing, and contributing to the overall well-being of residents. The temporary cycling network was created and aims to allow people on bikes to move around Toronto safely, to better connect the city, and to mirror major transit routes, providing key corridors. New cycle tracks, and bike lanes were rapidly installed through temporary materials, along with the Council-approved 2020 Cycling Network (Cycling Network Plan, Toronto). This is the largest one-year expansion of on-street bike lanes ever in Toronto, approximately 40 km of on-street cycling lanes approved for accelerated installation in 2020.

With the number of COVID-19 cases rising, and this new infrastructure in place, Canadians began to cycle or walk more than using public transport, as a safer alternative than crowded buses and subways, and perhaps diminishing the risk of getting COVID-19 as a result. Toronto also has a shared bicycle system, which covers more than 200 square kilometers. In 2020, Bike Share Toronto (2021) added e-bikes, to facilitate use for long-distance travel, and the number of users grew by 60% in the city. With private gyms and playgrounds temporarily closed, public outdoor spaces, such as parks, beaches, and squares, became vital to people's physical and mental health. There was a significant increase in the use of these outdoor spaces, despite the imposed lockdown. Painted signs were added to sidewalks with greater circulation to indicate social distancing. In parks, white physical distancing circles were painted on the grass inviting people to enjoy the outdoors while staying at a safe distance from each other (Fig. 3).

Restaurants in Toronto restaurants were closed for over one year since the start of the pandemic (March 2020), working only with delivery and/or people picking up merchandise purchased via telephone or internet. The City of Toronto created the

Fig. 3 Safety circles, Trinity Bellwoods Park, Toronto. *Photo* Isabella Trindade, June 2021

CaféTO (2021) program to guide and help these establishments to survive during the pandemic. The goal of the CaféTO program is to provide space for expanded dining areas to help restaurants and bars that are impacted by COVID-19, while maintaining physical distancing requirements, creating revenue generation, and bringing vibrancy back to our main streets.

The City of Toronto adopted a series of actions in support of these commercial establishments: (1) exemption from patio fees and expansion of these areas in public space; (2) relocation of public roads—spaces previously intended for parking, vehicle circulation to allow the creation and/or expansion of patios and parklets—to maximize the space of establishments, and thus better accommodate tables and chairs; and (3) establishment of guidelines, regulations, and fast permits to meet the required standards: public health, accessibility, fire safety, and road safety. This included six types of cafes: (1) Sidewalk Cafés, located on the sidewalk adjacent to the front; (2) Curb Lane Cafés, located in the curb lane of the road, in front of the restaurant; (3) Small Frontage Cafés, located directly against the building, with a maximum size of 0.8 m × 5.5 m or the width of the establishment frontage, whichever is less; (4) Flankage Cafés, located on the sidewalk or boulevard along the side of a restaurant, generally on a local street; (5) parking slot areas converted into public space with tables, chairs and food trucks; and (6) patios located on private property. In addition to CafeTo, Curb Lane Pedestrian Zones were designed to provide additional space for customer line-ups, and over 130 Temporary Parking Pick-up Zones in locations across Toronto (Fig. 4).

Fig. 4 (*Left*) Curb Lane café, Toronto. *Photo* Isabella Trindade, June 2021; (*right*) parking slot areas converted into cafés, Toronto. *Photo* Isabella Trindade, June 2021

3.2 Recife

The responses from the Brazilian government, which should have taken on a coordination role considering the National Health Service (*Sistema Único de Saúde*—SUS), were confusing and inefficient. Doubt and skepticism spread by the government headed by President Bolsonaro undermined the work of governors and mayors and contributed to increasing the number of cases and casualties. The Brazilian President called COVID-19 disease a "flu", disregarded social distance measures, did not wear a mask, and criticized the use of vaccines, thus inciting widespread doubt on their effectiveness. Governors and mayors had to produce their own regulations. Recife, located in the Northeast Region, is the fourth-largest urban area in Brazil, and like other Brazilian cities, suffered from a lack of public policies and informational guidelines at the national level. Unlike Toronto, where a lockdown was enacted and then an emergency urbanism began to be planned and implemented, in Recife, public authorities opted to fight the pandemic, at first, through extended social distancing, and the closure of parks, beaches, boardwalks, and bike lanes. Both the state and municipal governments regulated the functioning of services and activities, depending on the different stages of the pandemic, and the number of deaths caused by COVID-19. Thus, different degrees of greater or lesser flexibility for opening up commerce and services, and for the use of urban public spaces, were created depending on these fluctuating numbers. However, actions involving the practice of emergency urbanism were not implemented.

The controversies raised in Brazil by the different levels of political authorities (federal, provincial, and municipal) about the real dimensions of the spread of the virus corroborated the minimization of the severity of the disease caused by the virus, which gave rise to existing divergences among the population about the importance of, for example, the use of masks and/or vaccination. Recife's public space image thus reflected the actions, divergences, conflicts, and consequences arising from the pandemic and the economic crisis in the country. As in other places that underwent social isolation actions, deserted streets in areas of intense commerce and movement of people were configured as the main images of the city at the first moment of the pandemic. In Recife, the downtown areas were crowded with people. Many citizens in Recife did not observe the recommendations for the use of masks and social distancing. Thus, people without masks, clandestine parties, and crowded bars were in opposition to the recommendations of WHO and the public authorities. Public spaces as well as public transport became dangerous for their users in the face of the pandemic.

At the same time, planned actions of tactical urbanism were implemented in the city, promoting the widening of sidewalks and pedestrian crossings, through the painting of roads, and to better ensure walking and traffic management, contributing to urban connectivity. In addition, certain streets were presented as a space of social struggle and human rights, through the painting of pedestrian crossings, such as the colors of the rainbow, an inclusive message to the LGBTQIA + population (Fig. 5).

Fig. 5 (*Top left*) Intersection of Aurora St and Conde da Boa Vista Avenue (Rua da Aurora com a Avenida Conde da Boa Vista) (https://www.facebook.com/PTNUPORANGA13/posts/338428881221379)/; (*top right*) intersection of Ninfas St and Manuel Borba Avenue (Rua das Ninfas com a Av. Manoel Borba), Recife. *Source* Retrieved from http://www2.recife.pe.gov.br/noticias/26/06/2021/mais-seguranca-viaria-e-menos-preconceito-pcr-implanta-novas-areas-de-urbanismo. Accessed July 2021; (*bottom left*) Palma St before the sidewalk painting. *Source* https://www.google.com/maps. Access in May 2019; (*bottom right*) after the sidewalk painting. *Source* http://www2.recife.pe.gov.br/noticias/27/05/2021/pcr-entrega-rua-da-palma-requalificada-e-com-mais-espaco-para-os-pedestres. Access in May 2019

The painting of these streets in Recife, contributed both to the re-signification of the urban landscape, as well as to the practice of sustainable urbanism actions (see Farr 2013), since the streets accommodated more space for pedestrians and cyclists. Squares, beaches, and parks remained closed during the first stages of the pandemic and were re-opened in June 2021. Signs were placed in these locations to alert the population about the virus and prevention. Municipalities set up sidewalk signs for social distancing, mainly in front of the banks, due to the large number of people lining up to get help from government programs that were offering temporary financial assistance channeled through banks (Fig. 6).

Recife has bars, restaurants, and street vendors that traditionally use sidewalks as expanded dining areas to place tables and chairs for their customers, but no tactical urbanism actions focused on COVID-19 were identified. Bars and restaurants were closed at the beginning of the pandemic. Like parks and beaches, they were only partially opened, with 50% of the total capacity and social distancing protocols, depending on the different stages of the pandemic, and number of deaths caused by COVID-19. As for the use of non-motorized transport, Recife currently has 143.5 km of cycling routes, and 69 shared bike stations (Bike-Itaú program). Although new

Fig. 6 (*Left*) Signs at the Dona Lindu Park, Recife. *Photo* Lourdes Nóbrega, June 2021; (*right*) people lining up to get help from government programs. *Photo* Andrea Rego Barros.*Source* https://fotospublicas.com/prefeitura-do-recife-continua-com-acao-para-organizacao-das-filas-em-agencias-da-caixa/

emergency cycling routes that specifically responded to the demand of the pandemic were not created in the city as in Toronto, new cycling routes were implemented in response to the Recife Metropolitan Master Plan, which contributed to the increase of routes in the period analyzed. The City of Recife indicates an increase in the cycle route network—with 16 km completed in 2021.

3.3 New York City

The municipal authorities in New York City have launched several alternatives to deal with the restrictions regarding the use of public spaces imposed by the devastating effect of the COVID-19 pandemic on the city, where until September 2021, almost 55,000 deaths related to the SARS-CoV-2 virus had occurred (NYC HEALTH 2021). In this section, the three indicators of this study—circulation, permanence, and commerce—act simultaneously to stimulate the coexistence of people outdoors without compromising the fulfillment of supportive measures of social distancing, hygiene, and mental health, particularly needed in the pandemic context. The *Open Streets Program* created as a temporary initiative in April 2020 and made permanent in May 2021, established the closing of streets for vehicles, therefore making them safer for pedestrians and bikers who remained active and able to enjoy activities in their own neighborhoods. Allowing for the closing of streets to vehicles from 8am through 8 pm, by the end of 2020 the program had led to approximately 139 km (83 miles) of open streets (NYC DOT 2021a).

An instrument aiming to stimulate the circulation of people, it ultimately led to the use of sidewalks and deactivated parking spots (or parklets) as spaces for permanence and restaurant businesses. As stated by Carlina Rivera, a member of the New York City Council responsible for the program, this agile way of thinking should be carried over to transportation and infrastructure projects in the city, which could facilitate the post-pandemic recovery and the reconstruction of a greener New York City, at the same time more resilient and accessible to its residents (The City of New York 2021). The *NYC's Open Restaurant Program*, launched in July 2020, just after the *Open Streets Program*, aimed to expand the options of outdoor seating for restaurant businesses. The main goals were as follows: (i) to promote the use of open space; (ii) to assure social distancing (which could not be achieved indoors at the time); and (iii) to help businesses recover from the difficult economic context caused by the proliferation of the virus. In order to reach highly impacted neighborhoods, the program operated based on the principle of geographic equality, meaning the most affected areas were prioritized when it came to waiting times for approvals in the New York City Department of Transportation (NYC DOT 2021b).

The launching of the program coincided with the beginning of summer, when the city was about to reach Phase 2 of the reopening process which privileged a wider array of businesses, including food establishments that would benefit from being able to temporarily serve people outdoors. Remaining active to date, this initiative applied specifically to granting permits to install tables in the public space located immediately in front of the business. Regarding costs, the owners would be responsible for the installation of tables and chairs, as well as decks, guardrails, temporary tents, moveable ramps, among other items required for providing adequate service to customers.

Two subdivisions of the program were specifically related to restaurants: *Open Restaurants* and *Open Streets: Restaurants*. The first established parameters for individual businesses to utilize the sidewalk or the parking spaces in front of their storefronts. The latter determined requirements for community organizations (*Business Improvement Districts—BIDs*) or groups of three or more restaurants located within the same street block, in a collective partnership, to apply for a license to serve food outdoors in streets that would be temporarily shut down to car traffic. As a consequence, there would be two occupation typologies of public spaces: sidewalks and parking spaces (or parklets). Among the requirements for sidewalk seating, it is worth noting the following: the area of permanence could not exceed the commercial storefront; seating areas had to be at least 3 feet (90 cm) away from others; there should be a barrier-free pedestrian access measuring approximately 8 feet (2.45 m); waiting areas at bus stops should also be clear. These measures illustrate the concern about possible interferences of permanent space over those dedicated to pedestrian circulation in the city's public spaces.

Regarding restaurant seating, the following parameters are particularly relevant: to create a protection barrier along the entire perimeter of the seating area on the street, measuring at least 46 cm (18 in.) in width and 76–90 cm (30–36 in.) in height, which also preserves the drivers' visibility and assures customers' safety; to place these barriers directly adjacent to one another and no more than 2.45 m (8 feet) away from the curb; to provide an accessibility ramp, areas protected by umbrellas or tents should not exceed a total of 37m2 (398 sq ft) or go over the area limited by the barriers; and, lastly, the furniture must be removed (whether it is fixed or not) when the space is not in use. Therefore, it is evident from these measures that special attention is given to the condition of permanence of people in such spaces (Fig. 7).

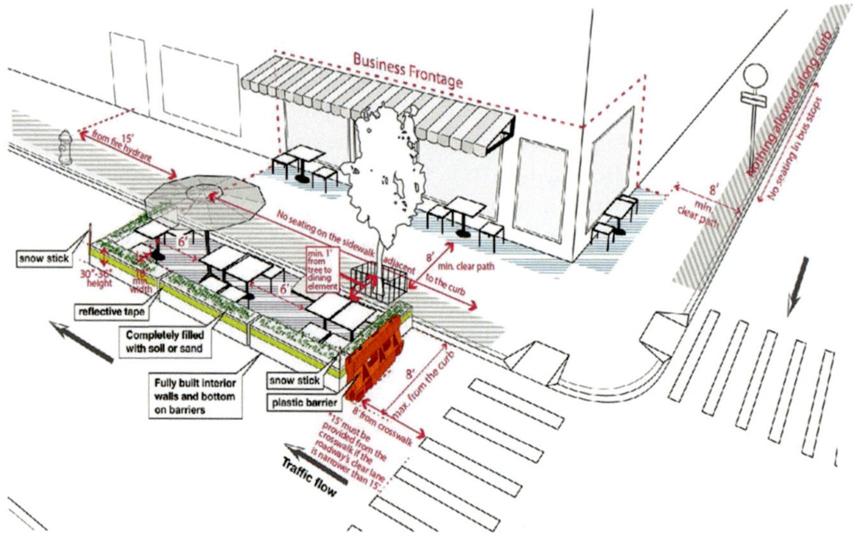

Fig. 7 Regulation for restaurant seating on sidewalks and parking spaces established by the *Open Restaurants* Program, New York City. *Source* https://www1.nyc.gov/html/dot/html/pedestrians/openrestaurants.shtml. Accessed on Sept. 5, 2020

One of the most relevant aspects of these programs is that, as a result of the emergence context imposed by the pandemic, the city created a normative instrument that was easily accessible to applicants and written in a straightforward manner, not requiring certified professionals (such as an architect or an engineer) to be involved in the approval process, configuring a sort of manual on how to intervene in public spaces. Thus, the initiative demonstrates that it is possible to simplify urban normative guidelines in an emergency context. This capacity to adapt and to act promptly connects directly to the concept of tactical urbanism, which implies the creation of temporary, bottom-up interventions that are capable of responding to the altered nature of public space, making cities more participatory and habitable (Uneven Growth 2014).

As an expansion of the *Open Streets: Restaurants* program, the *Open Boulevards Program* was launched in May 2021 to promote cultural activities, community events, landscaping, and other types of beautification, and art installations (NYC DOT 2021c). On such streets, small businesses, restaurants, artists, pedestrians, and cyclists could meet, creating the kind of destination expected in a city like New York. The priority was on the landscape and its use in the fight against climate change, in addition to expanding the offer of open public space. Currently, there are 15 *Open Boulevards* in the five neighborhoods of the city, which vary in length depending on the location.

Another recent program, effective since March 2021, is *Open Culture*, a type of license that allows the use of street lanes for performances controlled by tickets that are valid for one day. This encourages the recovery of the arts in the community, offering cultural experiences to the population (NYC DOT 2021d). Arts and culture are an important part of the efforts to improve the quality of life in the city, manifested in the works of art, performances, and cultural programs, which enhance the creation of attractive corridors, activating public space and expanding access to artistic, cultural, and educational resources. In this way, income generation opportunities are also created for these sectors.

Considering these alternatives, it is clear that despite the face of the "sick" context caused by COVID-19, a healthy city could emerge or as observed by the urban sociologist Sennett (1977), a healthy city that embraces and uses its lifestyles productively is different from the sick city, which isolates and separates the difference, without extracting collective strength from the mixture of different people. By facilitating the occupation of public space by users of different types of establishments, social classes, ethnicities, and genders, these programs in New York City have contributed to a vibrant, transitory, informal, and democratic public atmosphere.

In addition to mechanisms used to support businesses, the programs are safe alternatives for city dwellers to move, have fun, and eat outside their homes. By making a mix of different establishments and places accessible to the user, the principles of walkability are addressed. This characteristic has been observed at the scale of the neighborhood, where people can walk from their homes to the central streets, finding different activities that increase the liveliness, safety, and sociability of the place. True open-air dining rooms were created, in which strangers (separated by translucent panels or greenery) sit side by side, consuming in an expensive restaurant or a modest diner (Fig. 8).

Fig. 8 The program *Open Restaurants*, Chelsea, NYC. *Photo* Ana Luisa Rolim, Sept. 2020

Fig. 9 The program *Open Streets: Restaurants*, Chinatown, NYC. *Photo* Ana Luisa Rolim, Sept. 2020

Another result, especially with the *Open Restaurants*, was the fact that the lack of a licensed architect to sign the project for outdoor areas led the city to be occupied by objects conceived by non-specialists. This was similar to initiatives of placemaking where the use of few resources can result in significant transformations of urban spaces, which, in turn, became more user-friendly, even if produced in an improvised and temporary way. During this process, particular places benefited, such as the Chinatown neighborhood in Manhattan, where alternating seats and leisure areas were implemented in the streets. This gave the public space a rich atmosphere, which, before the pandemic, was restricted to the interiors of the picturesque restaurants in the neighborhood (Fig. 9).

4 Conclusions

A new form of "Emergency Urbanism" is revealing the promise of what the cities could be, as improvised "pilots" quickly implemented to expand the realm of the possible, testing what works and what is not minimizing economic and social losses for the government and for society as a whole. As this unfolds, the response reveals an entirely new way of using the city laced with interconnected safeways. This response is a powerful demonstration of city resilience, providing not only essential choices in a moment of crisis but also pointing the way to adaptations that have the great potential of becoming permanent. As we experience such urban change, the momentum is unlikely to be reversed, and can also create the city's opportunity.

The critical response to the current moment is the key to a safe and sustainable post-COVID-19 future.

In the three cases presented, it is noteworthy that planning, designing, and enabling the management and programming of public spaces, even in a moment of pandemic, directly contributed to the practices of sustainable urbanism, since, in this context, economic aspects and the physical and mental health of citizens were also considered. As life begins to return to normal, or to a "new normal", such initiatives constitute opportunities to assess the capacity for change and adaptation in the face of crises, representing creative alternatives for transforming urban spaces. They also provide an opportunity to think about the city we want, about how to live and overcome the deficiencies present in the urban environment, and toward achieving urban sustainability.

References

ActiveTO. Guidelines. City of Toronto. https://www.toronto.ca/home/covid-19/covid-19-protect-yourself-others/covid-19-reduce-virus-spread/covid-19-activeto/. Accessed August 2021

Bambozzi L (2019) Lugares e Espaços de informação. In: Bambozzi L, Portugal D (org.) O cinema e seus outros. Editora Equador, São Paulo

Bazant MZ, Bush JWM (2021) A guideline to limit indoor airborne transmission of COVID-19. https://www.pnas.org/content/118/17/e2018995118. Accessed August 2021

Bike Share Toronto (2021) 2020 expansion: Bike share Toronto grows to 6,850 Bikes & 625 stations. https://bikesharetoronto.com/news/expansion-2020/. Accessed August 2021

Bowser M (2020). Image cropped from the video posted by Washington Mayor Muriel Bowser, on June 5, 2020. https://twitter.com/murielbowser/status/1268916115809488896. Accessed July 2021

CafeTO. Guidelines. City of Toronto. https://www.toronto.ca/home/covid-19/covid-19-protect-yourself-others/covid-19-reduce-virus-spread/covid-19-cafeto/. Accessed August 2021

Cirella G, Bak M, Koźlak A, Pawlowska B, Borkowski P (2019) Transport innovations for elderly people. Res Transp Bus Manag 30:100381. https://doi.org/10.1016/j.rtbm.2019.100381

CurbTO. Guidelines. City of Toronto. https://www.toronto.ca/home/covid-19/covid-19-protect-yourself-others/covid-19-reduce-virus-spread/covid-19-curbto/. Accessed August 2021

Cycle Toronto. https://www.cycleto.ca. Accessed August 2021

COVID-19: Case Counts, City of Toronto. https://www.toronto.ca/home/covid-19/covid-19-latest-city-of-toronto-news/covid-19-pandemic-data/covid-19-weekday-status-of-cases-data/. Accessed August 2021

Farr D (2013) Urbanismo sustentável. Bookman Companhia Ed, Porto Alegre

Gehl J (2015) Cidades para pessoas. Editora Perspectiva, São Paulo

Greenberg K (2021) Five organizations come together to produce a community map of Toronto "Safeways". https://www.kengreenberg.ca/field-notes/five-organizations-come-together-to-produce-a-community-map-of-toronto-safeways. Accessed August 2021

Heemann J, Santiago PC, Guia do Espaço Público para Inspirar e Transformar. Project for Public Spaces: Nova York, S/D

Kahraman C, Orobello C, Cirella GT (2022) Changing dynamics with COVID-19: future outlook. In: Cirella GT (ed) Human settlements: urbanization, smart sector development, and future outlook. Springer, Singapore, pp 235–252

Malagi R, Mehta P, (2020) Tactical urbanism: an adaptive tool for safe distancing. Disponível em. https://thecityfix.com/blog/tactical-urbanism-adaptive-tool-safe-distancing-rajeev-malagi-prerna-mehta/. Accessed June 2021
Mídia Ninja (2020) Image retrieved from the Facebook page "Mídia Ninja" showing the street painting (in execution) of the movement "Vidas Pretas Importam" on November 21, 2020. https://www.facebook.com/MidiaNINJA/posts/2058170327674435. Accessed August 2021.
Moovit (2021) Moovit apresenta panorama de big data sobre transporte público em 10 grandes cidades brasileiras. https://moovit.com/pt/press-releases/relatorio-global-moovit-sobre-transporte-publico-2020/. Accessed August 2021
Nações Unidas Brasil (2020) A ONU e o meio ambiente. https://brasil.un.org/pt-br/91223-onu-e-o-meio-ambiente. Accessed July 2021
NYC DOT (2021a) Open streets. https://www1.nyc.gov/html/dot/downloads/pdf/open-streets-overview.pdf. Accessed 15 June 2021
NYC DOT (2021b) Permanent open restaurants program. https://www1.nyc.gov/html/dot/html/pedestrians/openrestaurants. Accessed 15 June 2021
NYC DOT (2021c) Open Boulevard program. https://www1.nyc.gov/html/dot/html/pedestrians/openboulevards.shtml#overview. Accessed 15 June 2021
NYC DO (2021d) Open culture. https://www1.nyc.gov/html/dot/html/pedestrians/dotart.shtml. Accessed 15 June 2021
NYC HEALTH. COVID-19: Data. Disponível em: https://www1.nyc.gov/site/doh/covid/covid-19-data.page. Accessed 16 Sept 2021
Organização Mundial da Saúde - OMS Brasil. https://www.who.int/eportuguese/countries/bra/pt/. Accessed August 2021
PlaNYC 2030. http://www.nyc.gov/html/planyc/downloads/pdf/publications/full_report_2007.pdf. Accessed August 2021
Qian H, Miao T, Liu L, Zheng X, Luo D, Li Y (2020). Indoor transmission of SARS-CoV-2. medRxiv, 2020.2004.2004.20053058. https://doi.org/10.1101/2020.04.04.20053058
Ravindra K, Singh T, Vardhan S, Shrivastava A, Singh S, Kumar P, Mor S (2022) COVID-19 pandemic: what can we learn for better air quality and human health? J Infect Public Health 15(2):187–198. https://doi.org/10.1016/J.JIPH.2021.12.001
Russo A, Cirella GT (2020) Urban sustainability: integrating ecology in city design and planning. In Cirella GT (ed) Sustainable human–nature relations: environmental scholarship, economic evaluation, urban strategies. Springer, pp 187–204. https://doi.org/10.1007/978-981-15-3049-4_10
Safeways TO (2021) https://www.cycleto.ca/safewaysto. Accessed August 2021
Santos M (1994) Técnica, espaço, tempo: globalização e meio técnico-científico-informacional. Hucitec, São Paulo
Senac SP (org.) (2018) Guia Global de Desenho de Ruas. Editora Senac, São Paulo
Sennett R (1977) The fall of public man. W.W. Norton & Co., New York
Speck J (2016) Cidade caminhável. Perspectiva, São Paulo
The City of New York. Streets Week!: Mayor de Blasio Makes Open Streets Permanent Part of New York City's Urban Landscape (2021a). https://www1.nyc.gov/office-of-the-mayor/news/361-21/streets-week-mayor-de-blasio-makes-open-streets-permanent-of-new-york-city-s-urban-landscape. Accessed 15 June 2021
Uneven Growth: Tactical urbanisms for expanding megacities (2014) The museum of modern art. Department of Publications, The Museum of Modern Art, New York
Vacinômetro. Prefeitura do Recife. https://conectarecife.recife.pe.gov.br/vacinometro/
World Health Organization (2021). https://www.who.int. Accessed August 2021

Informalized Workforce of Women Domestic Workers: Case of Bengaluru Metropolitan Region

Vinita Yadav and Sherin Jacob

Abstract A formalized workforce ensures economic stability, but the majority of the workforce engages in informal livelihoods. Professional women require support for which they engage women domestic workers (WDWs) hailing from marginalized and underprivileged sections. During the COVID-19 pandemic in India, such workers were the hardest hit. Middle-class and well-off sections managed to deal with the pandemic as the majority of them received salaries from their profit-making institutions while working from home, whereas WDWs were asked not to report for work to practise social distancing measures. They found it difficult to sustain their family due to the loss of income as fallout of stricter measures enforced to curtail the pandemic. The chapter aims at analysing the precarious socio-economic conditions faced by WDWs. The objectives are to analyse the causes accentuating the struggles of workers, their relationship dynamics with the employers and the role of stakeholders to balance the power relationship. The secondary data on the WDW's conditions was collected from published journal and newspaper articles. Data about the informal settlements have been collected from the publications of the Ministry of Housing and Urban Poverty Alleviation, the Government of India, the International Labour Organization and Participatory Research in Asia. The in-depth interviews using the snowball sampling technique were conducted with 5 WDWs in Bengaluru, to gather information on their socio-economic conditions in pre- and post-lockdown. The chapter concludes that WDWs, the providers of support to working professionals, are perceived as an obstacle during COVID-19. The multitudinous consequences of the lockdown impacted their socio-economic well-being. It recommends that capacity building is the way forward for ascertaining their rights and balancing the power dynamics. Civil society organizations shall play a proactive role to ensure their rights.

V. Yadav (✉)
Department of Regional Planning, School of Planning and Architecture, New Delhi, India
e-mail: v.yadav@spa.ac.in

S. Jacob
St. Claret College, Bengaluru, India
e-mail: sherin@claretcollege.edu.in

G. T. Cirella and B. Dahiya (eds.), *City Responses to Disruptions in 2020*, Advances in 21st Century Human Settlements, https://doi.org/10.1007/978-981-99-7988-2_5

Keywords Informal sector · Women Domestic Workers · COVID-19 pandemic · Unionization · India

1 Introduction

The advent of globalization opened up the Indian economy for global institutions, steering the country towards economic growth. These institutions established several subsidiaries creating job opportunities in the region. People's sustenance was dependent on the formal economy which provided permanent salaried jobs. High economic growth led to a demand for informal sector workers to be engaged in non-contractual jobs which inturn attracted large numbers of migrants from neighbouring districts (MoLE 2014). The informal sector refers to activities and economic units insufficiently covered under the formal sector. Out of the total workforce, 91% are informal workers contributing nearly fifty per cent of the total GDP in the Indian economy. They are primarily engaged in construction work, small scale-manufacturing units, home-based enterprises, traditional crafts, and household work (Murthy 2019). People from underprivileged sections form a significant portion of the informal sector's workforce, which is intrinsic to regional development. A major segment of the informal sector workers are women, who offer their services as domestic workers. Globally, 67 million people are engaged as domestic workers, and their numbers are increasing significantly in India owing to multiple social, economic and political factors (ILO 2021). Women domestic workers (WDW) work in an employer's household or multiple employers' households, on a part-time or full-time basis. They are aged between 18 and 75 years (NDWM 2023). The majority of them undertake gendered tasks and are disadvantaged in terms of access to economic opportunities, social protection and labour rights. This is despite the government mandating minimum wages for unskilled workers (HRW 2023).

The COVID-19 pandemic forced governments to implement lockdowns, across the world. Such lockdowns had devastating consequences for the vulnerable and marginalized population. Though the lockdowns intended to contain the virus, they resulted in large-scale unemployment amongst informal workers. Out of 116 million workers, 104 million informally employed workers lost their jobs in Phase 1 and Phase 2 of the lockdowns (Estupinan and Sharma 2020). Phase 1 began on 25 March and ended on 14 April 2020, whereas Phase 2 started on 15 April and ended on 3 May 2020. Though government provided support in the form of rations (free food), temporary shelters and free transport yet the strict containment measures, inadequate health infrastructure, precarious living conditions and loss of livelihood pushed them to the brink of poverty. COVID-19 highlighted the stark reality of the unequal employer–employee relationship (Reeves and Rothwell 2020). WDW, especially those living in large size metropolitan regions, namely Mumbai, Delhi and Bengaluru, were the hardest hit due to the announcement of the lockdowns. The pandemic exacerbated their challenges and difficulties as they were viewed as carriers of the virus due to their living conditions (Cirella 2020, 2022; Kahraman et al. 2022). The vulnerability of WDWs was owing to their position in the class and gender hierarchy which can be further understood with the help of social reproduction theory.

2 Literature Review

Informal labour is a significant component of the economy as it plays crucial role in the effective functioning of the capitalist economy. The lives of individuals, who comprise the informal labour force, should be viewed through multiple lenses and sociological theories as economic and social forces impinge upon their lives in multifaceted ways. Social reproduction theory (SRT) views and understands the social reproductive role of labour in capitalist societies. It refers to activities and institutions necessary for life-making in society and tasks instrumental for daily living (Jaffe 2020). Capitalism creates profit-making institutions that require a workforce (Bhattacharya 2017). Its proliferation has advantaged certain sections of society but disadvantaged others by facilitating gender and race oppression. SRT highlights the significance of the labour required to support this workforce. It accords prominence to institutions that nurture, sustain and create a labour force working for capitalist institutions. Formal workers earn higher wages which places them in a privileged position to hire informal workers, especially women domestic workers (WDW), at lower wages. Their labour enabled and freed the formal workforce from household work, but it is accorded lower value and considered inferior to other works. Their hard laborious work, despite being undertaken on a repetitive basis, is considered economically unviable in nature. Women are less interested in pursuing power-related goals as compared to men (McGinn and Oh 2017). When viewed from an intersectional lens, WDWs fall in the intersecting structures of caste, class and gender domains which impact the nature of work, work relationships, and influence their bargaining powers in the employer's homes (Mitra 2017; UN Women 2020). Such a lens recognizes the fault lines connecting the three domains and emphasizes upon the ways in which their overlapping nature disadvantages women (CIJ 2023; Massey 2022; Bernardino-Costa 2014).

Domestic work is considered invisible, unrecognized and unpaid work further contributing towards the marginalization of WDWs (Singh and Patnaik 2020). Domestic work is assumed to fall under women's domain as household-related labour is perceived as a natural extension of a woman's gendered role in society. Such workers are not recognized as 'workers' and their work is also 'undervalued' as they are mostly engaged in household-related work. Domestic work fails to come under the purview of a job requiring a special skill set. The absence of institutions like daycare facilities and surplus informal labour availability, places the employer in a better negotiating position. The latter increases the domestic workers' vulnerabilities as they can be fired at the whim of their employees. The absence of formal working conditions that guarantees job safety and security, increases their risk of being dismissed easily. This perception assigns the profession a devalued status affecting wage structure, working conditions, and employer–employee attitude (Basak 2022). Ghosh (2013) states that pre-existing social inequalities promulgat economic inequities for domestic workers seen as an intersection of caste, class, gender, and race. Such domestic workers are primarily poor, rural migrant, and lower-caste women (Mitra and Damle 2019). WDW are now increasingly drawn from the dominant caste due to a fall in rural income owing to agrarian distress.

Over the decades, domestic workers have mobilized and organized themselves into unions to address their burgeoning concerns with the support of secular and religion-affiliated non-government organizations (NGOs) and groups. Such unions fight against the socio-economic systems, state regulations, and civil society's related multiple oppressions. Such collectives devise strategies to deal with the discrimination experienced at the hands of their employers. They serve as the organizing ground for collective representation against the devaluation of their worth and subordinate position in society, owing to their intersecting identities (Marchetti, Cherubini and Geymonat 2021). Unions have trained and upskilled WDWs to advocate for their rights, ways to improve their bargaining and negotiation capabilities. The training enabled them to build solidarity and forge relationships to raise their voice against the injustice meted out to them by their employers. This has positively impacted the identity of WDWs as they started valuing their work and contribution towards the maintenance of employers' homes and overall society.

Against this backdrop, the chapter aims to analyse the socio-economic conditions, the impact of COVID-19 on WDWs and recommend strategies to uplift the lives of WDWs. The objectives are to examine the social and economic conditions of WDWs, relationship dynamics with the employers and other stakeholders and to formulate strategies for bringing equality and equity, asserting for their rights, and balancing power relationships. The chapter explores the interplay of caste and class, the role of Resident Welfare Associations (RWAs), the interrelationship amongst the workers and unions' contribution in empowering the WDWs. The difficulties experienced while tackling COVID-19 and its impact on their physical and mental well-being have also been analysed. Amongst migrant workers, WDWs' socio-economic condition is assessed in light of COVID-19.

3 Methodology

The literature on the topics of class dichotomy, wage structure, social reproductive labour and intersectional lens has been reviewed from Journals, books and websites. The information on social reproduction theory was obtained from books by international publishers, i.e., Pluto Press. Conceptual understanding of the Intersectional lens was sourced from the websites of UN Women, Center for Intersectional Justice (CIJ) and the blog of the London School of Economics. Conceptual understanding and information on socio-economic characteristics of WDW were obtained from international publications of Women in Informal Employment: Globalizing and

Organizing (WIEGO) and ILO. The newspapers, i.e., The Wire and The Hindu were reviewed to comprehend the pandemic's impact on WDW.

Secondary data about the total number of domestic workers, age, educational qualifications, employment and wage structure has been obtained from government websites and publications. The data relating to the wage structure of WDW was obtained from the Labour Commissioner Office of the government of Karnataka. ILO's document and the website of the National Domestic Workers Movement were referred for the filing of documents. The role of community especially living in gated areas and other stakeholders was reviewed from the websites of Outlook India and Citizen Matters. With the help of secondary data, measures undertaken to integrate WDWs and their contribution to a region have been referred to and analysed in the context of Indian cities. The secondary data, about the slums, has been collected from government documents namely Statistical Compendium on Slums in India from the Ministry of Housing and Urban Affairs (MoHUA). Op-edited articles from newspapers were referred to obtain secondary data on the WDW's conditions during the lockdown. It has also been gathered from publications of International Labour Organization (ILO) and Participatory Research in Asia (PRIA). The information about government schemes, i.e., *Ambedkar Karmika Sahaya Hastha Scheme*[1] (AKSHS) and *e-shram*[2] was accessed from the Ministry of Labour and Employment (MoLE), Government of India.

The primary data was gathered from the interviews conducted with members of the Karnataka Domestic Workers Forum as well as WDWs working in gated communities. The snowball sampling method was applied to identify the WDWs during the lockdown period. The snowball method is used by researchers to involve existing research participants to assist in locating other potential participants for a study. The contact details of registered WDWs were collected from the Union representative. During the lockdown, restrictions were imposed on the physical movement of the people. As everyone was confined to their homes, the researchers resorted to in-depth interviews as it was seen as the way forward to carry out the research. Indian Social Institute, Bengaluru was contacted to establish connect with Karnataka Domestic Workers Forum. Contact with the union leader, aged fifty-five years, was established to enrol participants for the study. The union leader provided the contact details of eight union members as potential participants. These union members were residing in Hennur and Horamavu wards in Bengaluru which hosts several Information Technology (IT) firms in the city. The members were informed about the nature and purpose of the research. Eight members were contacted over the phone for the research. Out of the eight members, five agreed to take part in the research, two members were unreachable and one potential participant refused citing personal reasons. Therefore, the number of respondents for the study was limited to five. The reference from the union leader proved helpful in establishing contact with the other respondents. The names of the respondents have not been revealed on request of anonymity by the interviewees. The in-depth interviews were conducted telephonically with the help of an open-ended questionnaire that elicited details from respondents on their demographic profile, employment status, nature of their job, the attitude of their employers during the lockdown period, the role of domestic workers union and socio-economic impact of job loss. Information on access to relief schemes was also sought to gauge the effectiveness of government schemes during the pandemic. Each of the interviews lasted between thirty to forty-five minutes per respondent. The insights shared by the respondents are representative of the larger universe. The number of respondents (5) for in-depth discussion may sound insignificant, but a thorough process has been followed to reach the respondents. The insights shared are representative of the larger universe. Information about the feelings, behaviour, situation and emotions of WDW especially during the pandemic has been derived from the primary data.

3.1 Bengaluru Metropolitan Region: Case Study

Bengaluru is famous for being the Silicon Valley of India. BMR is divided into 243 wards and around 8,443,675 persons reside in the region (Census of India 2011). Bengaluru, a global city, has witnessed rapid economic growth owing to the presence of Information Technology (I.T) companies and educational institutions. Being a host to companies, work avenues for white and blue-collared formal jobs have opened up. Formal sector employment has necessitated the services of informal sector workers. The informal sector workers, who hail from the surrounding villages, provide a host of services, i.e., construction work, street vending, and domestic work, thereby, contributing towards Bengaluru's growth. Hence, Bengaluru was chosen as a case to highlight the distress of informal sector workers.

The number of slums has grown exponentially in Bengaluru from 159 in 1971 to over 2000 in the year 2015. Out of approximate 8.5 million people, nearly 1.4 million reside in slums (PRIA 2014). Slum dwellers account for 25–35% of the Bengaluru Metropolitan Region (BMR) (Roy, Palavalli, Menon et al. 2018). As per Bruhat Bengaluru Municipal Palike (BBMP), 8.5% of slum households cater to 17% of the city population. Domestic workers are sought after and in high demand in working professionals' households. Their work frees the formal workforce from the drudgeries of household chores and enables them to undertake their formal jobs efficiently. COVID-19 pandemic severely impacted such workers, especially WDW, as they lost their jobs during the lockdown period. Thus, it is considered pertinent to study the WDW amongst informal workers to comprehend their distress.

4 Results and Discussion

Bengaluru's transition to a service-based economy required real estate developers to provide housing for the in-migrants. This inturn led to hiring of an informal workforce of carpenters, masons and other construction labourers. As per the Department of Skill Development, Entrepreneurship and Livelihood of Karnataka government report of 2016, Karnataka state employs nearly seventy-four percent of the workforce in the informal sectior (Ranganath 2019). Bengaluru hosts 4.2 million migrant labourers which accounts for 44% of the region's population. Many of the male members migrated for work with their families from neighbouring settlements. To survive in the metropolitan region, they allowed their better halves to search for jobs. The women's poor educational qualifications made it difficult to secure formal jobs, leaving them with the option of only searching for informal work and being absorbed in the informal workforce. A significant number of them undertook employment as domestic workers in gated communities and independent houses but they reside in informal settlements due to the absence of affordable planned spaces. They are assigned the tasks of cleaning, washing, cooking and taking care of children in middle and upper-class families (Bonnet et al. 2022). Their income was based on the nature of work, locality and household size of the workplace. The number of job seekers was more than the number of jobs available in the region. The availability of surplus labour diminishes their value and power to ascertain their rights. WDWs found themselves easily replaceable in the highly fragile and competitive labour market. This weakens their position and makes it difficult for them to escape the cycle of poverty.

During COVID-19, the female workforce experienced a significant decline in employment across various sectors. The impact of COVID-19 on the WDW was disproportionate when compared to the formal workforce employed in the capitalist economy. The latter was able to bear difficulties due to an assured income accrued from profit-making institutions, but it was not the case with the former.

4.1 Impact of Pandemic on Women's Choices: Experiences of Women Domestic Workers

Nearly 23% of Bengaluru's population, including WDW, reside in informal settlements owing to the non-availability of planned and affordable spaces in metropolitan regions (GoK 2023; Choudhry 2020). These settlements are situated on the encroached land originally earmarked for drainage, waste dumping and a green belt around industrial areas or below the high-tension wire (MoHUPA 2015). The occupants of such informal settlements lack tenurial rights which, further, impact their access to services. Living in such settlements prove precarious during natural and man-made calamities. Cramped living quarters located near unhygienic surroundings, coupled with the absence of basic sanitation services, made social distancing an impossible feat (Harish 2019). In the interest of public health, the government introduced measures to curb the virus spread around the third week of March 2020. The lockdowns brought the economy to a grinding halt. Though WDW understood the necessity of imposing the lockdown to safeguard their lives, it proved disastrous for their well-being. WDWs were unprepared to face a pandemic of such scale. All five WDWs were asked abruptly to stop reporting for work as it was non-contractual, precarious and outside the ambit of State regulation (Yadav and Jacob 2021). This made them jobless and attributed to the loss of steady monthly income. This was done with the prejudice of keeping the employers' family safe from the virus (OECD 2020). Prime Minister made a nationwide appeal to employers to pay the workers' dues to enable them to overcome the challenges. Despite this, except for one respondent, none of the remaining respondents received their wages. Inadequate financial reserves, in the form of savings, unleashed multiple hardships (Saluja 2022). The situation was further grim for single women-headed households. All respondents accepted the fact that they would not be paid at least till the crisis gets over. They were afraid to ask their employers about the remaining payment and the timeframe within which they would be called for work again. They expressed their dilemmas, feelings of helplessness and powerlessness while interacting with employers. When curbs were lifted in between the lockdowns, they became desperate to get back to their jobs. However, their rehiring did not pick up momentum due to fear of contracting the virus. Even when they were rehired, employers asked them to furnish documents such as police verification, *Aadhaar*[3], photograph and phone number. This, further, disadvantaged those WDWs who did not possess such documents.

For WDW, it was difficult to survive the lockdowns as loss of their livelihood made it difficult for them to pay their rent and other utility services. Their landlords threatened them to either pay or evict the home, leading to the fear of becoming homeless. Living in cramped dilapidated spaces, which are bereft of adequate water supply for maintaining hygiene, the possibilities of catching COVID-19 loomed large. One of the respondents wondered, 'How one could practice social distancing and good hygiene in slums (*excerpt of an interview conducted with a domestic worker on 30 June 2020*)?'.

During lockdown, WDW had to make a difficult choice between food and health. Four out of five respondents were the only breadwinners in the family. They expressed anguish at their inability to provide three meals a day to their children. All of them stated that they often remained hungry to save the ration which adversely affected their health. The steep fall in income impacted their access to health facilities which was of utmost concern for one of the respondents. In the words of one of the respondents from Bengaluru, '*she is sad due to her inability to continue health treatment for her intellectually challenged child. The public transportation in Bengaluru is affordable for the public but it is unavailable during a pandemic. This compounded her woes to access health services. As government pharmacies were closed, she was forced to purchase medicines from private shops which further made access to treatment difficult. At times, she has to forego her son's treatment in order to ensure food for the family (excerpt from the interview held on 29 June 2020)*'.

4.2 Role of Stakeholders in the Welfare of Women Domestic Workers

Structural inequalities cannot be addressed in societies where the ability to initiate strategic choices is constrained for women adhering to cultural value systems (Kabeer 1994) Individual revolts for autonomy are not considered positive, thereby, making such efforts unsustainable. Union has a proactive role in ascertaining the rights of the workers. Two of the respondents were union leaders. They were well-versed in issues about domestic workers due to their knowledge of relevant Acts. These respondents said affirmatively that their association with union had enhanced their knowledge about their rights. They were outspoken about their challenges, but their forthrightness was of limited help. Both union leaders shared similar experiences. However, it had put them in uncomfortable situations with their employers. The employers often questioned them as they did not place the WDWs Union work in high regard. A union leader stated that '*she is often ridiculed because of her association with the union. She was instructed to do "Law" related work outside the gamut of employer's house. She was criticised when she raised critical questions about her rights. She felt stifled working in such unjust conditions. Despite being aware of the unfairness meted, she decided to bear the situation and not raise her voice to safeguard her family's well-being and sustenance (excerpt from the interview held on 29 June 2020)*'. Her gender determined her position in the power dynamics between an employer and employee relationship. They were forced to think about whether union or domestic work had greater primacy. To quote a respondent, '*Are we doing our domestic work or doing this law related work? (excerpt from the interview held on 29 June 2020)*'. WDWs are still reluctant to initiate independent acts of collective mobilization, challenge and transform different forms of material and social injustices in their lives (Kabeer 1999, 2012). Unions act as a bridge between employers and employees. According to Barua and Haukanes (2019), an association with domestic workers' union tends to bring about an attitudinal shift in the way women perceive cognitive changes in terms of empowerment, knowledge enrichment and awareness of their entitlements. It sensitizes the resident welfare associations (RWAs) and households towards WDWs.

RWA organized a social benefit camp to enable WDWs access to government schemes. The camp was organized for workers working within their jurisdiction. To support the informal sector workers, Prime Minister launched the *Pradhan Mantri Jan-Dhan Yojana*[4] under which INR 500 was deposited in their account. Only one out of five respondents reported that she received the amount, but it was insufficient to meet the expenses. The government distributed relief materials to support the vulnerable section of the population. It was the only mean for multiple families to survive during COVID-19. Yet, the distribution of relief material had political undertones. A respondent shared that, '*they received mainly sugar, rice and lentils. The ration distributed in their area was insufficient, of poor quality, contaminated and had less nutritive value. The food was unable to build immunity due to poor nutrient content. This took a toll on their physical health. The local powerful leaders usurped the relief material which was supposed to be distributed in their locality. Thus, it did not reach the families who needed it most. For her, the imposition of lockdowns should have preceded with a consideration of its impact on the vulnerable section (excerpt from the interview held on 30 June 2020)'.* Lack of jobs, along with poor savings, to fight a raging pandemic took a devastating toll upon their physical and psychological well-being. COVID-19 reversed significant strides made with respect to gender equality and women empowerment.

The government of India has implemented the welfare schemes of *Ambedkar Karmika Sahaya Hastha Scheme* (AKSHS) and *e-Shram.* Under the AKSHS scheme, launched by Karnataka State Unorganized Worker Social Security Board, Bengaluru, eleven categories of informal workers are identified and issued *Karmika*[5] (Worker) smart cards. The scheme provided domestic workers an identity and individuality in the informal sector. Ministry of Labour and Employment, GoI has launched the *e-shram* portal to provide centralized data of informal workers, including WDWs in India. The portal enrols, registers and identifies workers and links them with stakeholders to provide the benefits of social security schemes. National Skill Development Corporation (NSDC) is a not-for-profit public limited company with forty-nine per cent share of the company with GoI-Ministry of Skill Development and Entrepreneurship. The aim of NSDC is to provide skill-based training to workers.

WDW juggles between production and social reproduction activities such activities restrict opportunities for leisure and social mobilization. It inhibits WDWs' participation in social movements, which inturn hinders their structural, economic and social transformation. The absence from the workplace entails a pay cut, which makes it difficult to participate in union work. Devaluation of self and work is a hurdle in motivating WDWs to carry out both i.e. their job and their quest for seeking justice through union work. They have less time at their disposal to advocate for their rights at the workplace. There is a reluctance to organize, mobilize and work collectively to enhance their socio-economic status due to fear of losing the job. This affects their abilities to question the employers and exercise control over their own lives. Such situations force them to desist from exercising their rights and negotiate for higher wages, which further pushes them into deeper vulnerabilities. Such challenges raise a key question: *'How can the WDWs empower themselves when they are fraught with existential and survival needs?'.*

5 Conclusions and Recommendations

The pandemic made upper and middle-class households realize the contribution of WDWs. It brought back attention towards the concerns of WDWs, working in hostile and stigmatized environments. The pre-existing structural class, caste and gender inequalities were amplified due to socio-economic conditions related challenges. WDWs' existentialism came under threat due to the perception of their employers. They faced multitudinous consequences of the lockdowns, including denial of their basic rights. The crisis re-iterated the need to adopt a positive outlook and a multi-pronged approach to address the concerns of WDWs with the active involvement of government and civil society members.

The pandemic gave a wake-up call to the government and other stakeholders to be prepared for any such events in future. Resident Welfare Associations (RWAs) shall be sensitized towards the plight and concern of WDW. They shall ask the residents to adopt the uniform wage rate as per the work, number of family members and duration of engagement. There is a need to bring an attitudinal shift in the employers' outlook towards domestic workers. They shall be sensitivised to understand their interdependence and guarantee fixed-term employment on mutually conducive terms. Efforts shall be taken to make informal sector workers understand the importance of formal work and assert their rights to get formalized. The employer and employee relationship shall be based on an egalitarian outlook.

The government launched many of the welfare schemes for the intended beneficiaries, but workers were not found aware of such schemes. Beneficiaries need to be linked with social and economic schemes of the government. Unorganized workers should be encouraged to enrol on the *e-Shram* portal. National Skill Development Corporation (NSDC) shall use such information to enrol WDWs as trainees. The training will enable them to work as entrepreneurs and lessen their vulnerability to economic shocks. They should be provided loans to start up their enterprises along side working as WDW. This will empower them economically and improve their self-worth. The unions should encourage the WDWs to enrol for the women empowerment schemes of the government. This will provide them an opportunity to be alternatively employed in future, if need arises. This will also minimize their vulnerability and enable them to get out of the trappings of the cycle of poverty and discrimination. The union shall be assigned the task of registering the WDWs and issuing '*karmika* cards' under the AKSHS scheme. This is to acknowledge their identity as workers and receive the benefits of government schemes.

Women domestic workers shall be empowered to secure their financial stability by asserting their rights. Capacity-building measures shall be introduced to generate awareness about various schemes and their benefits. Their capacity shall be built to interact, bargain and assert their rights. This, along with the provision of social and financial safety nets, will enable them to exercise their personal choices and claim their entitlements. Their socio-economic development will remove survival concerns for which network building along with social and family support is a must. Domestic workers union shall invite WDWs and their family members for network building. Union shall provide the advocacy support to WDWs. This will help them to overcome the pandemic-related feeling of disempowerment and abandonment. The collective efforts will ensure the regularity of income. Such mutual support will help the women leaders and members to further build upon their network and contribute to union-related work. Government should recognize the unions and assign them a role in building capacity and ensuring the rights of its members.

Endnotes

1. Ambedkar Karmika Sahaya Hastha Scheme registers informal workers and issues smart card to recognize their services.
2. e-Shram Portal stores details of informal workers in a centralized database, which serves as a reference point to link the informal workers with job opportunities as well as enable them to receive the benefits of welfare schemes.
3. Aadhar is a unique identification number provided to the citizens of India.
4. Pradhan Mantri Jan-Dhan Yojana (scheme) was launched to ensure the access of financial services, i.e., credit and pension to the informal workers.
5. Karmika (workers) smart card has been issued to eleven categories of unorganized workers including Domestic Workers.

References

Barua P, Haukanes H (2019) Organizing for empowerment: exploring the impact of unionization on domestic workers in India. Stud Int Comp Dev 55(1):27–47

Basak S (2022) Blame-game always begins with us: woes of domestic workers in Noida's gated societies, September 26. Outlook India. https://www.outlookindia.com/national/blame-of-theft-always-lands-on-us-woes-of-domestic-helpers-in-noida-gated-societies-news-225790. Accessed 25 Apr 2023

Bernardino-Costa J (2014) Intersectionality and female domestic workers' unions in Brazil. Women's Stud Int Forum 46:72–80. https://doi.org/10.1016/j.wsif.2014.01.004

Bhattacharya T (ed) (2017) Social reproduction theory: remapping class, recentering oppression. Pluto Press, London

Bonnet F, Carré F, Vanek J (2022) Domestic workers in the world: a statistical profile. https://www.wiego.org/publications/domestic-workers-world-statistical-profile. Accessed 16 Apr 2023

Census of India (2011) Bangalore (Bengaluru) city population 2011–2021 (2021). https://www.census2011.co.in/census/city/448-bangalore.html. Accessed 28 Dec 2022

Center for Intersectional Justice (CIJ 2023). What is intersectionality? https://www.intersectionaljustice.org/what-is-intersectionality. Accessed 16 Apr 2023

Cirella GT (2020) Sustainable human-nature relations: environmental scholarship, economic evaluation, urban strategies. Springer, Singapore

Cirella GT (2022) Human settlements: urbanization, smart sector development, and future outlook. Springer, Singapore

Choudhry V, Avinandan V (2020) Why COVID-19 outbreak in India's slums will be disastrous for theurban poor. https://www.outlookindia.com/website/story/opinion-covid-19-outbreak-in-indias-slums-will-be-disastrous-for-theurban-poor/350335. Accessed 8 Apr 2020

Estupinan X, Sharma M (2020) Job and wage losses in informal sector due to the COVID-19 lockdown measures in India. https://ssrn.com/abstract=3680379. Accessed 25 Aug 2022

Ghosh J (2013) The plight of domestic workers in India. http://www.macroscan.org/cur/jan13/pdf/Domestic_Workers.pdf. Accessed 20 Dec 2022

Government of Karnataka (GoK 2023) Abstract details of slums. Slum Development Board. https://ksdb.karnataka.gov.in/page/Slums/Abstract+Details+Of+Slums/en. Accessed 25 Apr 2023

Harish S (2019) Urban development, housing and slums. In: Patel S, Goyal O (eds) India's contemporary urban conundrum. Routledge, Abingdon, Oxon, pp. 184–198

Human Rights Watch (HRW 2023) Domestic workers. https://ndwm.org/domestic-workers/. Accessed 16 Apr 2023

ILO (2021). Distinguishing the concepts: the informal sector, informal employment and the informal economy. https://www.ilo.org/global/topics/wages/minimum-wages/beneficiaries/WCMS_436492/lang--en/index.htm. Accessed 15 Jan 2021

Jaffe A (2020) Social reproduction theory and the socialist horizon: Work, power and political strategy. Pluto Press, London, UK

Kabeer N (1994) Reversed realities: gender hierarchies in development thought. Verso, London and New York

Kabeer N (1999) Resources, agency, and achievements: reflections on the measurement of women's empowerment. Dev Chang 30:435–464. https://doi.org/10.1111/1467-7660.00125

Kabeer N (2012) Women's economic empowerment and inclusive growth: labour markets and enterprise development. Discussion Paper No. 29. Centre for Development Policy & Research, School of Oriental & African Studies, University of London.

Kahraman C, Orobello C, Cirella GT (2022) Changing dynamics with COVID-19: future outlook. In: Cirella GT (ed) Human settlements: urbanization, smart sector development, and future outlook. Springer, Singapore, pp 235–252

Marchetti S, Cherubini D, Geymonat GG (2021) Global domestic workers: intersectional inequalities and struggles for rights, 1st edn. Bristol University Press. https://doi.org/10.2307/j.ctv209xn9k

Massey S (2022) Designing social policies through an intersectional lens, March 15. London School of Economics. https://blogs.lse.ac.uk/socialpolicy/2022/03/15/designing-social-policies-through-an-intersectional-lens/. Accessed 25 Apr 2022

McGinn KL, Oh E (2017) Gender, social class, and women's employment. Special issue on inequality and social class. Curr Opin Psychol 18:84–88

Ministry of Housing and Urban Poverty Alleviation (MoHUPA) (2015) Slums in India. A statistical compendium. http://nbo.nic.in/pdf/SLUMS_IN_INDIA_Slum_Compendium_2015_English.pdf. Accessed 16 Apr 2023

Ministry of Labour and Employment (MoLE 2014) Employment in informal sector and conditions of informal employment. https://labour.gov.in/sites/default/files/Report%20vol%204%20final.pdf. Accessed 25 Apr 2022

Mitra S (2017) How the Noida incident shows the feudal face of Indian cities, April 17. Citizen Matters. https://citizenmatters.in/noida-domestic-workers-riot-feudal-city-codes-4595. Accessed 25 Apr 2022

Mitra R, Damle A (2019) Female migrant workers and domestic employees need a security net, November 11. The Wire. https://thewire.in/labour/women-labour-domestic-work.

Murthy SV, Ramana (2019) Measuring informal economy in India-Indian experience. International Monetary Fund. https://www.imf.org/~/media/Files/Conferences/2019/7th-statistics-forum/session-ii-Murthy.ashx. Accessed 16 Apr 2023

National Domestic Workers Movement (NDWM 2023) Domestic workers. https://ndwm.org/domestic-workers/. Accessed 16 Apr 2023

OECD (2020) Youth and COVID-19: Response, recovery and resilience. Organisation for economic co-operation and development. https://read.oecd-ilibrary.org/view/Youth-and-COVID-19-Response-Recovery-and-Resilience. Accessed 14 Aug 2020

Participatory Research in Asia (PRIA 2014) Government led exclusion of the urban poor: a greater contributor through a lesser recipient, Bangalore Study Report 2014. PRIA, New Delhi

Ranganath C (2019) When finding a job became a nightmare in Bengaluru after note ban, GST. https://www.business-standard.com/article/economy-policy/when-finding-a-job-became-a-nightmare-in-bengaluru-after-note-ban-gst-119050200125_1.html. Accessed 16 Apr 2023

Reeves R, Rothwell J (2020) Class and COVID: how the less affluent face double risks, 27 March 2020. https://www.brookings.edu/articles/class-and-covid-how-the-less-affluent-face-double-risks/. Accessed 16 Apr 2023

Roy D, Palavalli B, Menon N, King R, Pfeffer K, Lees M, Sloot P (2018) Survey-based socio-economic data from slums in Bangalore, India. Sci Data 5(1):1–9. https://doi.org/10.1038/sdata.2017.200

Saluja R (2022) How COVID-19 worsened hardships of India's domestic workers, 11 May 2022. https://thewire.in/rights/how-covid-19-has-worsened-hardships-of-indias-domestic-workers. Accessed 16 Apr 2023

Singh P, Patnaik F (2020) Unfolding unpaid domestic work in India: women's constraints, choices, and career. Palgrave Commun 6(1):1–13. https://doi.org/10.1057/s41599-020-0488-2

UN Women (2020). Intersectional feminism: what it means and why it matters right now. https://www.unwomen.org/en/news/stories/2020/6/explainer-intersectional-feminism-what-it-means-and-why-it-matters. Accessed 16 Apr 2023

Yadav V, Jacob S (2021) Impact of the lockdown on women domestic workers. Indian J Public Adm 67(3):300–313. https://doi.org/10.1177/00195561211044530

Critique of the Indonesian Omnibus Law Regime: Misguided Interpretation in Narrating "Village Owned Enterprises"

Aurora Meliala, Anis Rifai, and Jonathan Andre Woods

Abstract The COVID-19 pandemic led the Indonesian government to initiate the Omnibus Law, which has impacted all Indonesian legal frameworks, specifically in regard to economic law. As mentioned in the Governmental Regulation Number 11 of 2021, one breakthrough that has been proposed by the Government of Indonesia to accelerate economic growth is the narrative of Village-Owned Enterprise (or *BUMDes*). However, the regulation has become a polemic as it has deviated from the essence of Indonesia's economic democracy. While the Constitution of Indonesia has clearly mandated the development of Indonesia's economy through cooperativism, the government is instead focused on capital accumulation in the development of *BUMDes*. This chapter sheds light on the concept of *BUMDes* and how the village's economic independence can be achieved through law. This research article expounds on the subject in question in seven consecutive narratives. First, it explains the general background of the omnibus in lieu of law. It then expounds on the sudden shift that occurs in village economic policy because of the economic shock introduced by the COVID-19 pandemic. It then explains the original design of economic democracy in Indonesia's constitution. The fourth part elaborates on the structure of village-owned enterprises (*BUMDes*) as shaped by relevant regulations while the fifth part explains *BUMDes*' status as a legal entity and a relevant problem regarding it. The last part presents cooperatives as a healthy alternative to and a recommended framework for *BUMDes*. The research finds that *BUMDes* was constructed to be similar to a limited company and is focused on the capital dimension and concluded that a reflection on cooperatives as values and institutional ideas is needed to construct *BUMDes* as an economic tool truly based on cooperativism.

A. Meliala (✉) · J. A. Woods
Faculty of Law, Universitas Pembangunan Nasional Veteran Jakarta, Depok 12450, Indonesia
e-mail: aurora@upnvj.ac.id

J. A. Woods
e-mail: jonathanandrew@upnvj.ac.id

A. Rifai
Faculty of Law, Universitas Al-Azhar Indonesia, Jakarta 12110, Indonesia
e-mail: anis.rifai@uai.ac.id

G. T. Cirella and B. Dahiya (eds.), *City Responses to Disruptions in 2020*, Advances in 21st Century Human Settlements, https://doi.org/10.1007/978-981-99-7988-2_6

Keywords Cooperatives · People-centered economy · Indonesian economic law · United Nations sustainable development goals

1 Introduction

The omnibus concept was first introduced to the Indonesian legal system by President Joko Widodo on 20 October of 2020 and was applied to the Law Number 11 of 2020 on Job Creation, which was promulgated to overcome the impact of uncertain global economy because of the COVID-19 Pandemic (Prasetyo and Setiadi 2023). The law contains hundreds of articles that govern multiple sectors, from labor to the environment. Many of its provisions are made to smoothen economic undertakings. Of interest for this chapter is its article 117 which declared that village-owned enterprises are legal entities which brought more certainty for investors and stakeholders. However, the law attracted much criticism and various disagreements from workers, activists, and civil society organizations, especially in terms of labor rights.

As a result of this, on November 4th, 2021, the Constitutional Court declared the Law on Job Creation contrary to the constitution. In Decision No. 91/PUU-XVIII/2020, the court ordered the parliament to revise the law within two years, otherwise, it will become permanently unconstitutional and unenforceable. Instead of following up on the court's dictations, the government made a sudden maneuver on New Year's Eve and introduced the Government Regulation in Lieu of Law Number 2 of 2022 (omnibus in lieu law), a replacement with practically the same provisions. This decision also requires the Government of Indonesia to suspend any strategic and broad-impact measures or policies that are based on the law, as well as to cease the issuance of its implementing regulations.

This tireless endeavor for this omnibus law originated in light of the current condition of Indonesian legislation. Many deemed the system hyper-regulated, overlapping, and fraught with difficulties for the private sector.[1] This perplexed legislature and time-consuming law procedures had put Indonesia in the 91st position out of 190 countries in the World Bank's assessment of "the ease of business" in 2017. As has been mentioned previously, the omnibus in lieu law was made to smoothen many economic undertakings. In its wake, Indonesia's regulatory landscape is clearly set to be dominated by regulations aimed at maintaining economic growth. It particularly pivoted on the improvement of the investment ecosystem and its key breakthroughs were the deregulation and debureaucratization in many sectors. Attracting more foreign direct investment (FDI) is seen as a high priority to help fuel post-COVID-19 economic recovery and growth (Kahraman et al. 2022). Despite this, the law still lacks the impact that it promised, perhaps bordering on futile, as Indonesia still ranked low in the Corruption Perception Index as it scored only 34 from 100 and

[1] There are 40,988 total regulations consisting of 3841 Central Regulations, 16674 Ministerial Regulations, 4491 Non-Ministerial Government Agencies Regulations and 15982 Regional Regulations.

placed 110 out of 180 countries surveyed (Transparency International 2022). The debureaucratization seemed inconsequential. An inquiry into the effects of the law became imminent. The law simplification strategy of the omnibus in lieu law may not be as fruitful as it was first thought in helping the state accelerate its economy.

To attain a more complete portrait on this matter, the current legislation on village economy is now also put into question with the subject of discussion being the village-owned enterprises (abbreviated *BUMDes* in Indonesian) as it was championed as the main economic undertaking of a village unit. The Omnibus in lieu law has produced two implementing regulations in regard to *BUMDes*, the Governmental Regulation Number 11 of 2021 on Village-Owned Enterprises and Regulation of the Minister of Villages, Development of Disadvantaged Regions, and Transmigration Number 3 of 2021 on the Registration, Data Collection and Ranking, Guidance and Development, and Procurement of Goods and/or Services for Village-Owned Enterprises/Joint Village-Owned Enterprises.

The original economic ideals of Indonesia's founding fathers, such as Mohammad Hatta, that was enshrined in article 33 of the 1945 Indonesian Constitution preferred a resilient national economy that must be built upon robust economic ideology and people empowerment. Cooperativism is the center of such an economy along with an inherent social character. The high level of dependence on capital according to Hatta, would make a country susceptible to global crises (Hatta 1975). A shift from this narrative was acutely indicated in the omnibus in lieu law. Where the economic ideology mandated cooperativism, the law introduced *BUMDes* which was principled on corporatist practices.

There are indications that lawmakers attempted to incorporate constitutional ideals into *BUMDes* regulations, but they are arguably paradoxical in nature. For example, while Article 87 of Law Number 6 of 2014 concerning Village and the omnibus in lieu law regulated that *BUMDes* is managed with a familial spirit and mutual cooperation and that it can operate in the public services sectors, implementing regulations enacted only postulate these ideas narrowly. The Governmental Regulation Number 11 of 2021 did not accentuate the importance of social character as it was the case with cooperativism.

Several past research in the practice of *BUMDes* reveal an element that adds to this problem. Amanda (2017), Sumiasih (2019), and Utami, Tripalupi, and Meitriana (2019) have reported on the positive benefits that *BUMDes* brought to villages. The Ministry of Villages, Underdeveloped Areas, and Transmigration (Ministry of Villages) is only focused on structuring the management of village economy with the *BUMDes* instead of reforming it. Utami et al. (2019) further noted that while *BUMDes* economic management does contribute an improvement to a village's income, that improvement cannot be directly felt by the community. Moreover, advantages that *BUMDes* brought to the village's economy can only be observed limitedly as previous research often only provided local empirical evidence of one village or another. Many *BUMDes* are also only in their nascent stages. The novel and not well-documented nature of BUMDes implementation in Indonesia has also prevented definitive research to be conducted in regards to its impact towards Indonesia's gross domestic product (GDP) or the like. To date, no national data is available to determine their definitive contribution to the national economy. Thus, abstracting a conclusion on *BUMDes* implementation from such fragmented evidence may be misleading. Dirkareshza et al. (n.d) informed that *BUMDes* implementation even indicates ineffectiveness due to several factors, namely (i) segregated and disharmonize regulation; (ii) less contribution from rural community; (iii) management unaccountability; (iv) limited information and market access; etc.

Instead of addressing these problems according to Indonesian ideals, the Ministry of Villages recently decided to sign a Memorandum of Understanding with the Ministry of Investment to encourage the collaboration of investors in village economy by investing in the *BUMDes*. Government Regulation Number 11 of 2021 also decided to revitalize village-owned joint ventures (*BUMDesma*), a *BUMDes* created by two or more villages, as a companion to *BUMDes*. This *BUMDes*-alternative recently received a more comprehensive treatment than the previous regulation in the Governmental Regulation Number 11 of 2021 which further enables its creation alongside *BUMDes*. However, without a resolution to the basic issues which already trouble *BUMDes*, the actual advantage of the use of these policy tools is easily questionable.

This chapter provides a comprehensive examination of seven interconnected themes pertaining to the governance of *BUMDes* (Village-Owned Enterprises) in Indonesia. Firstly, it explores the abrupt transformation of village economic policies resulting from the introduction of the omnibus in lieu law in response to the COVID-19 pandemic. Subsequently, it delves into Indonesia's inherent economic ideology, encompassing cooperative principles and their ideal implementation within *BUMDes*. The chapter proceeds to analyze the existing BUMDes regulations, highlighting their characteristics and potential drawbacks. It then focuses on the legal entity of *BUMDes*, addressing associated challenges and concluding the analysis of the current *BUMDes* regime by highlighting a significant issue: the incompatibility between current *BUMDes* regulations and Indonesia's cooperative economy. Finally, the chapter underscores the cooperative model as an exemplary, people-centered economic institution.

2 The Shift of Village Economic Policy After the COVID-19 Pandemic

The revitalization of *BUMDes* in the Job Creation Law advances an alternative to cooperatives in Indonesia's modern legal landscape. As it is, it mirrors past instances of disaster capitalism that already happened in Indonesia. One occurred during the fall of Soekarno, Indonesia's first president, in 1966 when the communistic practice of the state was overturned by an increasingly capitalist one under one General Soeharto, not long before the first systematic regulations on investment were introduced (Klein 2008). Another can be said to have transpired during the COVID-19 Pandemic which saw the introduction of the Law on Job Creation, an investment-advancing omnibus law, in 2020. Recent research has presented a concrete view of this current era's shock therapy, where the chaotic episode of the pandemic was utilized by governments to introduce emergency packages, reset power, and rearrange governance relations.

This "governance shock doctrine" initiated a political strategy which consolidated government interests with those of their apparatus and affiliates (McGee 2022). This doctrine was especially obvious in the economy sector which was the most affected policy space with a negative impact doubling the 2008 financial crisis (Clemente-Suárez et al. 2021). Using the reason to curb the growing uncertainty of the resulting socio-economic upheaval, governments have enacted policies (Klein 2008) which provided immense relief for the private sector and further commodification of economic resources.

For example, in the United States, the Trump Administration had ordered the Environmental Protection Agency (EPA) to suspend its enforcement of environmental laws and lowered fuel emission standards for locally produced vehicles. Following this, India introduced its 20 trillion-rupee revival and policies on privatization as a way to accelerate the post-pandemic economy. Governments were thus charading in the creation of abrupt measures as a way to "survive" the pandemic even though the enactment of such policies may not cater to public interest (Mdalose et al. 2020).

The reason for economic uncertainty and shock was also used in the creation of the Indonesian Job Creation Law. The government reasoned that the act was constructed to combat the underlying economic threats that emerged in light of the pandemic (Sundari and Amalia 2020). When the omnibus in lieu law which replaced the problematic law was enacted in 2022, the same reason was once again used with a reason that the law is needed to counteract the "most dire situation" the state is currently in (Hipan and Budahu 2023). The evolution of this particular omnibus law where policies are abruptly enacted in a "state of emergency" to provide a seemingly more reliable economic narrative amidst socio-economic troubles fully portrays what is termed a "shock therapy" (Burdon 2022).

The shock instigated by the pandemic along with the momentum of the omnibus in lieu law was also used by the government to introduce a more fleshed-out *BUMDes* to Indonesia's legislature. While the regime itself was not anywhere new as *BUMDes* was already regulated by Law Number 6 of 2014 on Villages and its implementing Governmental Regulation Number 43 of 2014, its ruling provisions differed quite substantially from the new regime enacted by Governmental Regulation Number 11 of 2021 on *BUMDes*.

One substantial difference between them was on how expansive *BUMDes* were being regulated. The pre-omnibus in lieu of law regulations briefly legislated *BUMDes*. Many of its provisions were worded generally with the necessary details for its implementation, such as its particular duties, being relegated to either village regulation or in the *BUMDes*' articles of association. The new regulation introduced a more comprehensive but also more restrictive regime towards *BUMDes*. The institutionalization of *BUMDes* was thus framed by the central government which presents a more restricted policy space for the village to regulate *BUMDes*.

Another substantial difference between the pre- and post-omnibus in lieu law regulations was on the matter of *BUMDes*' entity. In Regulation Number 43 of 2014,

BUMDes was regarded as a business entity only and not a legal one. The explanation of article 87 (2) of the regulation justified this identification by substantiating that the *BUMDes* cannot be regarded as a legal entity alongside cooperatives and limited liability corporations. *BUMDes* was in its own class as a business entity, aimed and proportionally restricted at the village economy. This was abruptly changed in Regulation Number 11 of 2021. In article 1 (1) and the general explanation of the regulation, *BUMDes* was made a legal entity which required it to adjust to common corporate principles. Under the new law, *BUMDes* is now mandated to be registered to the Ministry of Law and Human Rights, become a subject to taxes, and submit reports to the government that detailed its performance (Simamora et al. 2022). While this may increase the availability of *BUMDes* to receive investments, as investors become increasingly assured of its legal status and its rights and obligations under the law, the *BUMDes* may become unprepared and inadequate in implementing this change as there was hardly any coordinated assistance to do so from the central government, which then negates the supposed impact this amendment targeted to make.

Adding to such an abrupt change was also the time accorded to *BUMDes* to adjust to the new regulation under Governmental Regulation Number 11 of 2020. Article 75 (1) of the regulation stated that the *BUMDes* that already existed before the government regulation came into effect, are required to adjust to the new provisions of the regulation no later than 1 (one) year after the regulation was promulgated. The time accorded by the regulation for the implementation of the new regime was relatively short, considering how many *BUMDes* were already created and continue to be established each year.[2] Implementing regulations also accorded the *BUMDes* with many new obligations to be fulfilled outside the reporting system that *BUMDes* must adopt since it is recognized as a legal entity. These issues undoubtedly add to the sense of abrupt change that the economic-legal shock as introduced by the omnibus in lieu law demanded.

3 Understanding the Design of Indonesia's Economy Democracy

Understanding the ground realities of economic activities is the key to developing good policy and regulation. This is true, as the foundation of economic development is dependent on social norms (Eriksson 2015). Mohammad Hatta, the father of Indonesian economic thinking, stated long ago that Indonesia's economy must be rooted in the idea of people's development and must not rely only on capital.

[2] According to the Ministry of Villages, Development of Disadvantaged Regions, and Transmigration, there are 47,500 BUMDes in Indonesia at present, most of which are registered online. The Government of Indonesia aims to increase the number of BUMDes by another 10,000 by 2024. It is hoped that this will consequently support economic growth, widen employment opportunities, and eventually decrease the urbanization rate (Pemerintah targetkan jumlah BUMDes berkembang capai 10.000 pada 2024—ANTARA News n.d.).

Following this line of thinking, BUMDes must rely on its social capital which covers the motivation of the rural community to give more contribution for village development. Social capital is the ideal fundamental basis as well as a source of energy in the development of village-owned enterprises. Social capital can be manifested not only through people's participation in investment but also in trust and willingness to work (Dolfriandra Huruta et al. n.d.). If we refer to the consideration of BUMDes establishment, it is further explained that it must have a social mission, meaning that BUMDes should represent social entrepreneurship.

The narrative of social economy should not be purely economic but also social in the sense that it also serves the interests of those that partake in a particular economic undertaking. Indonesia's social economy can be found in its *Ekonomi Rakyat* (Indonesian socialism) which emerged at the time of the economic depression in the United States during the 1930s. During that time, Hatta wrote about the miseries of Indonesian people under the past regime of *Cultuurstelsel* and expressed his concern in several declamatory articles. Two most influential articles on this subject were "*Pengaroeh Koloniaal Kapitaal di Indonesia*" (or "The Impact of Colonial Capital in Indonesia" in English) and "*Ekonomi Ra'jat*" (or "People-Centered Economy" in English). There are at least three main ideas that were delivered through the articles:

1. Hatta rejected the idea of national dependence on colonial capital. He strongly criticized the premise stating that the economy of Indonesia was dependent on the capital injected by colonial powers and/or foreign investors because of its contribution to national tax revenue.
2. Hatta emphasized that the decrease of state's stock was not only solely related to *belasting* (tax) but also to the two other variables: *invoerrecht* (customs duty) and *handel* (trade income). The domination of foreign direct investment and/or colonial capital would make a country susceptible to global crises.
3. Therefore, the resilience of a national economy must be built by a robust ideology of *Ekonomi Rakyat* (Indonesian socialism).

The distinction between *Ekonomi Rakyat* (Indonesian socialism) and the two poles of prevalent world economic ideologies is described in the Table 1.

By crystallizing these principles of a "people-centered economy," Hatta has introduced the idea of *Ekonomi Kerakyatan* ("People-centered Economy"). Indonesian economic policy along with its law products according to Hatta has to be focused on labor-intensive programs rather than capital-intensive ones. He asserted that economic activity must not be centralized only in big industrialized cities but must be motorized in cooperation between the city and the village. Hatta institutionalized this idea in the 1945 Indonesian Constitution. The soul of Indonesia's economic democracy in the constitution was proscribed in the constitution's article 33 which placed the interests of society as first and foremost. This is portrayed in the elucidation of the article: "*...In economic democracy, the priority is the prosperity of society not the prosperity of the people ... Prosperity for all people...*".

Table 1 Distinction between neoclassical economics, socialism, and *Ekonomi Kerakyatan*

Neoliberalism—Capitalism	Socialism	*Ekonomi Kerakyatan* ("People-centered Economy', according to Constitution of Indonesia)
Based on the principle of personal freedom/liberalism	Based on the principle of rights and responsibilities	Based on the principle of mutualism
Individualism	State/civil society	Fraternity/brotherhood
Competitiveness	Universalism/pragmatic mix of universalism and selectivity	Cooperativism
Market-sovereignty	Market-sovereignty	People-sovereignty
Homo-economicus	Civilian/part of community	Homo-humanus Homo-socius Homo-ethicus Homo-religious Homo-magnificus
Self-interest	State	Mutual-interest
Capital-centered	Authority-centered	People-centered, people-based
Maximum individual profits	Equality of outcome/ employability	Social welfare
Exploitation economics	Equitable economic access	Democracy economics
Individual business ownership	High cash/asset redistribution	Public business ownership
Social ideology is considered as caritas and/or philanthropy	Social Ideology is defined by the state or considered limited only to pragmatic benefit such as national minimum wage	Social ideology embodied social civil rights

Source Meliala (2022)

Article 33 (1) principally declared that *"the economy is structured as a joint effort based on the familyhood principle, and the suitable company structure is cooperatives."* That is to say that Indonesia's economy must be constructed based on the communal undertaking and be built on the collective endeavor of the people. Moreover, the economy should not be bestowed upon the market but upon the people with the state playing a pivotal role in protecting the people's prioritization. To accomplish this, article 33 (2) was given the ammunition to control production factors that are important for the livelihood of the people as it regulated that: *"... any production that is important to the State and affects the livelihood of the people are* ***controlled by the state*** *..."* The people's position is central-substantial and should be prioritized as further explained by article 33 (3): *"... Earth and water and natural resources contained therein are controlled by the State and used for the* ***greatest prosperity of the people*** *..."* Abstracting from the above-mentioned explanation, the concept of *Ekonomi Kerakyatan* ("People-centered Economy"), as a "middle way" ideology, was initiated in response to the failure and incompatibility of communism and liberalism. The concept is embodied in Article 33 of the 1945 Constitution.

Furthermore, the structure of *Ekonomi Rakyat* demands the placement of cooperatives in the Indonesian economy which is its paramount manifestation. Considering the place of cooperatives in the system of Indonesia's *Ekonomi Kerakyatan*, its elaboration is usually divided in two complementary narratives. The first being cooperative as an institution and the second being cooperative as a value set within an economic system. Within the context of this chapter, the second meaning is most appropriate to be elaborated. Cooperative as a value set in an economic system is the structure of the whole economy that is built on the cooperative spirit. In this second meaning, cooperatives are the pillar (*sokoguru*) of the Indonesian economy as stated in the Elucidation of Article 33 of the 1945 Constitution.

Cooperatives have become the most promising catalysts to reform the structure of the Indonesian economy that was encumbered by the predominant issue of poverty. Cooperatives are seen as the most eligible institutions that can provide welfare as well as justice in economic access. Hatta (1954) further noted that in cooperatives, there are no employers and workers. In it, all workers jointly endeavor towards common interests based on the familyhood principle and in accordance with the values and culture of the Indonesian nation.

The resemblance that the concept of cooperatives has with the character of Indonesia's economic democracy that is centered in kinship and people empowerment ensures that cooperatives can easily grow and develop. Swasono (2018) explained that the concentration of economic power which potentially builds economic dominance is especially precluded under Indonesia's democracy. This preclusion, as Hatta put it, is to be realized by cooperatives as the primary tools of social justice, especially for the lowest economic groups even in rural areas (Kahin 1980). Analyzing the contents of Hatta's ideas reveal that they are undeniably rooted in the character of kinship in rural areas that was upheld by almost all village communities. Thus, Indonesian cooperatives truly reflect the collectivist value which is rooted in the original customs of Indonesian life. Summarizing the points above, *Ekonomi Kerakyatan* was built upon empirical customs such as the habit of mutual assistance (*tolong-menolong*) and communal work (*gotong-royong*) that was and is truly present in Indonesia's community life (Hatta 1971).

Witnessing the development of today's economics and how the concept of Cooperative might fit with the structure of BUMDes, especially how the village character is clearly embedded in the framework of cooperative, several adaptations can be made to ensure that BUMDes is fundamentally sound. Firstly, the values of cooperatives must be incorporated into the principles of BUMDes itself. This includes the economic-social character that the original cooperatives have. Incorporating this character means that the profits of *BUMDes*, apart from being available to be economically cultivated, can be used directly for the betterment of the village community. This comes in the form of a direct distribution of profit to village members and which constitutes the fulfillment of *BUMDes*' responsibility as a social institution. Another value is joint cooperation. As Hatta has expounded, the cooperative system entails no hierarchy in its day-to-day undertaking and that all components strive for the same common ideal. *BUMDes* must enable itself to realize this in a sense that all its stakeholders can strive for the betterment of the village without being compartmentalized by hierarchical positions.

Further aspirations of Hatta can be conceptualized in the *BUMDes* were expounded by the Indonesian economist, Sri-Edi Swasono. According to him, the Indonesian economy can still be catalyzed by cooperative values with the fulfillment of its Triple-Co characterization (Swasono 2018). These three-fold pragmatic values consist of **co-ownership**, **co-determination**, and **co-responsibility**. These three principles feature the ultimate constitutional mandate in carrying out economic activities by all and for all ("togetherness") under the leadership or supervision of community members. In the context of *BUMDes*, co-ownership is realized when there is joint ownership between *BUMDes* management and village citizens. Co-determination entails that all stakeholders including village citizens are involved in determining *BUMDes* policies. While co-responsibility necessitates the joint responsibility of all stakeholders including village citizens in all *BUMDes*-related matters. The application of the Triple-Co characterization in *BUMDes* arguably provides the necessary togetherness to all its stakeholders that the village economy has as its original foundation. Communal empowerment in this cooperative scheme will, in the end, bring a natural entrepreneurship to the village community that is required for a resilient local economy envisioned by Hatta.

4 The Structure of Village-Owned Enterprises: *BUMDes*

It is important to highlight that Indonesia's democracy not only relied on its political freedom—access to the political elections—but also on its economic democracy where all economic resources are accessible to the society. This is slightly different from the goal that is narrated on the *BUMDes* regulations which only aimed at welfare. *Ekonomi Kerakyatan* went beyond the notion of a production-based economy for the benefit of the people which required a big amount of capital and must be profit-generating. Its ultimate aim is to build an economy where the **people are producers.** The ultimate goal is to empower the community to be, at least, economically independent.

Moreover, such an objective is easier to realize at the village level as it has an advantage in terms of direct and unadulterated access to natural resources in its area. Money capital will not be the only factor in the village's advancement if the village can also economically utilize those natural resources. However, the biggest obstacle in realizing such a potential is the direly needed optimal human resources. Focusing only on an attempt to advance the village economy through *BUMDes* without the educational establishment of resilient manpower to actually ensure *BUMDes*' sustainability can be regarded as a misleading tactic. Indeed, the omnibus in lieu law and its implementing regulations have no provision in regard to the quality of human resources needed to run *BUMDes*. Without sufficient human resources, the social capital as mentioned by Huruta et al., (n.d.) as being fundamental in providing motivation for rural communities to willingly contribute to village development cannot be realized.

Nevertheless, Law Number 6 of 2014 considered the economic independence of villages as the *BUMDes*' ultimate goal. Besides human resources, institution building, and its related mechanisms is the pivotal aspect of *BUMDes*, considering the robustness of the institutional framework needed for Indonesia's complex economic geography. The village is the basic unit of autonomy in Indonesia's governance and a catalysator for economic distribution. However, developing village economics requires a multifaceted approach and BUMDes regulations are far away from realizing such independence in rural economics. Villages are commonly known for their low purchasing power and poverty, and this was also burdened by previous policies such as one enacted by the Law Number 6 of 2014 which accorded only one billion rupiahs as a fund for the village system. This low-budget situation was what was utilized by legislative and presidential candidates to, indicatively, turn BUMDes into a political sale used to seek public sympathy in the general election (*Pengamat: Dana Desa Rp 1 Miliar Jadi "Jualan Politik,"* n.d.).

BUMDes is considered as an independent business institution that can sustainably provide the welfare to the village community (Ridlwan 2013) with the village administration being indicative of its success. BUMDes is prescribed to be embedded in the implementation of government affairs and the interests of the local community in the government system of the Republic Indonesia. There are several elements of village administration as described on the Law Number 6 of 2014, among others:

a. the Village Head (Article 1 point 3 of Law Number 6 of 2014)
b. the Village Consultative Body (Article 1 point 4)
c. Consultation Forum that involves the village community (Article 69 paragraph (9)

The incorporation of BUMDes has to be narrated under Village Regulation which has been discussed and consulted first with the village community. However, the mechanism of consultation is not explained further in Law Number 6 of 2014 and also Government Regulation Number 43 of 2014 concerning Implementation Regulations of Law Number 6 of 2014 concerning Villages. The consultation forum is recently described as *MusDes (Musyawarah Desa)* under The Government Regulation Number 11 of 2021. Nevertheless, it still does not touch on how exactly a decision is to be made within this specific forum and what are its specific requirements. Following Indonesia's constitution, for example, which has high standards on the qualification of *Musyawarah*, an Indonesian style deliberation procedure based on traditional customary law, will require full consensus *(mufakat)* of all stakeholders which may be challenging to achieve.

When narrated, the political motive for BUMDes is aimed not only to create income but also a strong social vision. For example, BUMDes in waste management must not only be projected to create profit but also it has to be able to answer or solve rural residents' problems regarding waste and the creation of a clean environment. BUMDes is expected to increase rural economic growth since it has a significant role to be in charge from upstream activities to downstream. It can help the government in distributing subsidies in agriculture so that fertilizer and seeds become more affordable to farmers. Concurrently, BUMDes can also become both the aggregator of rural production and the trustworthy financial service provider who can not only manage credit but also raise funds and even manage account payables/receivables.

However, by analyzing the architecture of BUMDes on its related legal products as can be seen briefly in Fig. 1, the structure of BUMDes has several similarities to State Owned Enterprises. Nevertheless, the conditions of BUMDes legal entity is not further described. Under what type of legal entity should it be established, is it a *Perseroan Terbatas*—limited liability company or a cooperatives institution? Recent discourse narrates that the BUMDes will be formed similarly to BUMN (State Owned Enterprises) which is bound to the regulation of limited liability companies. Thereby basically the legal form of BUMDes is aimed for the objectives of capital intensiveness, not the labor one. It is projected to be developed as joint ventures and not joint efforts. The BUMDes authorized capital will be fully sourced from the village treasury, where the management of the village treasury is entirely in the hands of the village apparatus (in this case, the Village Head). With this structure, the Village Head will become the full ruler and manager of the BUMDes. Village communities are not positioned as owners, controllers, or managers of BUMDes.

Having analyzed the structure of BUMDes as regulated under Law No.6 year 2014 in conjunction with Government Regulation No.1 year 2021, BUMDes is closely related to the limited company legal entity, the comparison is as follows (Table 2).

Furthermore, the analysis of articles of association of several *BUMDes* especially on profit sharing indicates, and arguably confirmed, that *BUMDes* is more akin to a capital association—capital intensive—than a social entrepreneurship of village citizens. For example, article 23 of the articles of association of *BUMDes Lancar Barokah* owned by the Joho Gunung Village, Rembang divided the *BUMDes*' business profit between addition for capital, reserved funds, fees for the management units, and village revenue. There is silence on the use of revenues for the direct benefit of individual villagers and instead, focused on the accruement or piling up of capital (Table 3).

Thus, BUMDes can be concluded to be similar to State Owned Enterprises (BUMN) and Regional Owned Enterprises (BUMD). They are a business entity that provides goods and services to earn a profit. The difference is that the capital in BUMD is dominantly supported by the local government. Profits received by BUMD or Regional Companies (PD) are one of the elements in Regional Original Management (PAD), while BUMDes are directed to contribute to the improvement of village revenue sources (PADes). The social initiatives are yet further elaborated on the BUMDes regulations. It seems the terms of social and local wisdom have been put only as an artifice in consideration to make such legal products look noble.

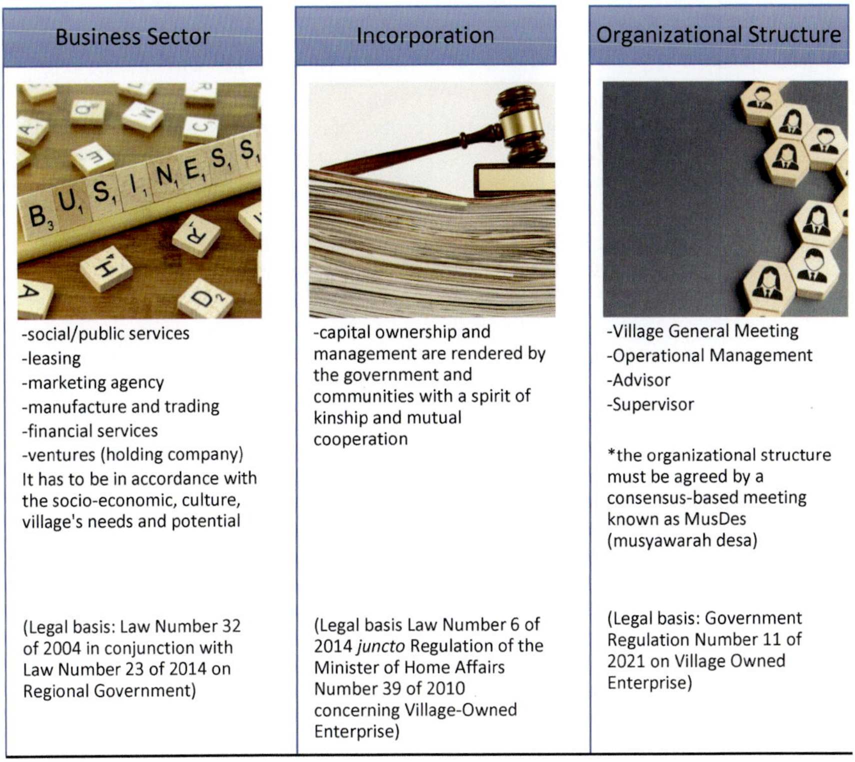

Fig. 1 The architecture of *BUMDes*. *Source* Law Number 32 of 2004, Law Number 6 of 2014, and Governmental Regulation Number 11 of 2021

4.1 Status of and Problem with BUMDes as a Legal Entity and Elaboration of a Fault in Current BUMDes Regime

The acknowledgment of *BUMDes* as a legal entity has been stipulated in the Governmental Regulation Number 11 of 2021, especially its article 1(1) and general explanation. Previous research on the legal entity of *BUMDes* such as done by Simamora, Ester, and Kristy (2022) determined that recent governmental regulation has led to a legal implication where *BUMDes* has the following characteristics:

(1) Separated assets;
(2) Have a specific purpose;
(3) Have their own interests; and
(4) There is an organized organization.

Moreover, current regulation has determined that *BUMDes* is not parallel to other legal entities such as a limited liability company, or a cooperative. It is a whole new business entity that has been legitimized by the omnibus in lieu law which also provides convenience for *BUMDes* to run businesses, even though in a previous

Table 2 The difference between limited company and *BUMDes*

Limited company (PT)	Village-owned enterprises (BUMDes)
It is based on legal entity—there's a responsibility separation between the allocated shares and the private asset *(Legal basis: Article 1 point 1 Law 40 year 2007)*	BUMDes is also incorporated based on legal entity that is specifically aimed at commercial activities and public services *(Legal basis: Article 1 point 1 Government Regulation No 11 year 2021)*
Capital is raised by the allocation of financial and/or tangible assets and/or intellectual property *(Legal basis: Article 34 Law 40 year 2007)*	Capital is raised by the allocation of financial and/or tangible assets *(Legal basis: Article 41 Government Regulation No 11 year 2021)*
Operational is controlled by the director; supervised by commissary and should be held accountable in annual or extraordinary General Meeting *(Legal basis: Article 92, 108, and 75 Law 40 year 2007)*	Operational is managed by the director; supervisor; head village as an advisor; and should be reported annually or occasionally in Musyawarah Desa (MusDes) *(Legal basis: Article 16, 21, 24, 28 Government Regulation No 11 year 2021)*
Share-holders are eligible to gain dividends equivalent to the number of shares *(Legal basis: Article 171 Law 40 year 2007)*	The terminology of Work Rewards (Hasil Usaha) is defined as business income minus expenditure within one annual financial report. Further, the Work Rewards is distributed equivalently to the amount of shared capital *(Legal basis: Article 60 Government Regulation No 11 year 2021)*

Source Law Number 40 of 2007 concerning the Limited Company and Governmental Regulation Number 11 of 2020 on *Village-Owned Enterprises*

Table 3 Difference between the profit sharing of the *BUMDes* and the cooperative

BUMDes	Cooperative
The use of the business profit is as follows: a. Addition of capital b. Village income c. Advisor cost d. Supervisory body cost e. Operational cost f. Education and social endeavor cost g. Reserve *(Based on article 23 of the Articles of Association of BUMDes Lancar Barokah at Joho Gung Village)*	The remaining profit is distributed to: a. Members according to their merits b. Members according to their savings c. Capital reserves d. Management and supervisory funds e. Social fund (*Based on article 23 of the Articles of Association of Melayu Raya Cooperative of Sabuh Village)*

Source Articles of Association of BUMDes Lancar Barokah and Melayu Raya Cooperative

section of this research, regulations have implied that *BUMDes* is principally similar to a "limited company."

However, there is quite an issue with the status of *BUMDes*, in its capacity as a legal entity, as having separated assets. Originally, article 1(6) of the Law Number 6 of 2014 on Villages determined that the assets of *BUMDe*s are separated from the village's own. This definition was retained in the Governmental Regulation Number 43 of 2014. Furthermore, the condition of separated assets in *BUMDes* was further confirmed in article 135 (3) of both Governmental Regulation Numbers 43 of 2014 and 47 of 2015. Herawati (2016) summarized that the wealth of BUMDes is the wealth of the separated village. BUMDes also has the objectives and interests set out by law, namely to develop the village economy and increase village income. BUMDes also has an organized organization which can be seen from the presence of advisors and operational implementation.

However, this matter becomes complex following a list of happenstance:

- The Governmental Regulation Number 11 of 2021 has revoked article (135) (3) in article 76. With this the definition that *BUMDes* has separated assets depended on article 1 (6) of the Law Number 6 of 2014 and article 1 (7) of Governmental Regulation Number 43 of 2014.
- Article 1 (1) of Governmental Regulation Number 11 of 2021 did not state that *BUMDes* has separated assets. Moreover, nowhere in the governmental regulation did it clearly state that *BUMDes* has separated assets.
- While article 77 of the Governmental Regulation Number 11 of 2021 stated that previous regulations governing *BUMDes* are still in place, the previous *BUMDes* regime regulated it as a business entity, while the newer governmental regulation regulated it as a legal entity. Thus, the lack of synchronization in the complete definition of BUMDes appears as a fault in regulation.
- The delineation that *BUMDes* as a legal entity has separated assets is currently based either on its definitions under previous regulations or on its definition as a legal term by law scholars.

This perceived ambiguity can become a major concern once the new regime is completely in place. In the case that the lack of an article regulating *BUMDes* having separated assets can be interpreted, in a rough manner, as it having unseparated ones, this potentially endangers the assets of the village as a whole. Because BUMDes' assets originated from the village's treasury, it is possible that when payments for debts with the BUMDes' assets are insufficient, creditors can demand village assets for further payments. This issue must be resolved quickly with, at the very least, adding one or several articles that clearly determine that *BUMDes* has separated assets in Governmental Regulation Number 11 of 2021.

Departing from this problem on *BUMDes* as a legal entity and abstracting from the analysis of this research particularly on the current regime of *BUMDes*, there is one exemplar issue that *BUMDes* experienced which further proved it as being not very compatible with the cooperativist system of Indonesia's Economy. A problem in regards to profit sharing. The right to regulate profit sharing, as mandated by article 60 (3) of Governmental Regulation Number 11 of 2021, was given to be regulated thoroughly in articles of association of a *BUMDes*. However, research that went into the implementation of profit sharing in such articles of association proved that *BUMDes* failed to realize the social function that is inherent in the cooperative economic system. *BUMDes* Profit distribution did not provide village members a chance to directly receive *BUMDes* profit as is possible with cooperatives.

Furthermore, the profit that is being given to the village is directly made village income. Despite article 60 (2) ensuring that village income use can be determined by the *MusDes*, it may not stop village officials from taking charge of the village income and unlawfully using it for their own gains. Such a case already happened in Indonesia with an example being the corruption of a village head of Dassok, Pamekasan, East Java where the village income was misused to bribe a government official. In order to eliminate this flaw and the threat that may arise from it is imperative for the government to include a provision in the governmental regulation that enables direct profit sharing for village citizens.

4.2 Cooperatives as an Alternative for BUMDes

After considering *BUMDes* and its structure, governing regulations, and faults, the article will now consider the cooperative institution as its alternative in fulfilling the cooperative ideals that already existed in traditional Indonesian values. Cooperatives as economic institutions originated in Europe and began to be developed in Indonesia in 1896 (Hasan et al. 2018). Hatta was known as the father of Indonesian cooperatives and their inclusion in the 1945 Constitution, at the start of Indonesia's independence, was supported by many. One such supporter was Soepomo, who was known as the architect of the constitution (Saifudin 1996) who thought that "*the mutual help system, characterized through cooperation, should be used as one of the economic bases of the Indonesian State*" (Soepomo speech at the BPUPKI session on May 31, 1945).

The definition and structure of cooperatives in the Indonesian Cooperatives Law has a long and complex evolution. It was first legitimized under Law Number 25 of 1992 with a history of an attempted amendment to update its regime in Law Number 17 of 2012. The definition of cooperatives under the 2012 regulation clashed fundamentally with its definition according to the International Co-operative Alliance (ICA). ICA defined a cooperative as "an autonomous association of persons united voluntarily to meet their common economic, social and cultural needs and aspirations through a jointly owned and democratically-controlled enterprise" (ICA 2015). It is important to highlight that the distinction of Cooperatives is based on its association of people. However, the 2012 regulation, instead of understanding a cooperative as an "association of persons," defined it merely as a "legal entity," which emphasized not its identity and fundamental nature but its required legalization, through administrative registration by the state (Pulungan 2021).

The 2012 Indonesian Cooperatives Law only lasted about one year before the Constitutional Court canceled it with Decree number 28/PUU-XI/2013. In evaluating a number of articles that regulate the norms of cooperative legal entities, members' shares, the authority of the supervisory board, and the cooperative board, the court found that the regime offered by the law has revoked the core principles of Indonesia's economic democracy such as people's sovereignty, principle of kinship, and togetherness as mandated by the 1945 Constitution (*UU Perkoperasian Dibatalkan Karena Berjiwa Korporasi* n.d.).

The definition of a cooperative as a "legal entity" was criticized as the root problem. The 2012 regulation potentially enabled the corporatization of cooperatives and expanded the opportunities for capital participation by non-members. This form of capital participation can be regarded as a fundamental threat to a cooperative's independence. Moreover, the Constitutional Court considered that Article 1 (1) of the 2012 regulation which stated that cooperatives are legal entities did not contain a substantive meaning and can even be referred to as a tyrannical company incorporation. This is not in line with the nature of cooperatives as referred to in Article 33 (1) of the 1945 Constitution.

Concluding from the narrative above, it can be said that when the form of legal entity potentially dispels the key point of people as a core structure of cooperatives and replaces it with a capital alliance or accumulation, the title of Cooperatives itself will no longer be relevant. In the end, the Constitutional Court decided that the definition of Indonesian cooperatives as explained under article 3 of Law No. 12 of 1967 and article 1 (1) of Law Number 25 of 1992 prevailed. Thus, Indonesian cooperatives can be safely defined as people's economic organizations that are characterized by social motives and consisting of individuals or legal entities which constitute an economic structure as a joint effort based on the principle of kinship.

Outside of this legal problem, Cooperatives have been proven to be effective agents in fostering democratic knowledge and practices without social exclusion. They have shown resilience in the face of economic crises. Its existence is well-placed within and can contribute to the triple bottom lines of economic, social, and environmental objectives in sustainable development because while they are enterprises that endeavor to meet the economic progress of their members, they also satisfy prevailing socio-cultural interests and protect the environment (ILO 2015).

Considering the problem mentioned above, *BUMDes* have fundamental problems in its regime when it is faced with the institutional framework of Cooperatives. First is the prospect of *BUMDes* being a legal entity. Law Number 25 of 1992 on Cooperatives enabled cooperatives to receive a status as a legal entity as mandated by article 1 (1). But it did not prevent it from prioritizing the individual prosperity of its members, who partake in the realization of triple-co principles to this very day. The current *BUMDes* regime was given the status of a legal entity in a framework that encouraged a capital-intensive concentration instead of a labor-intensive one according to Indonesia's ideals will only benefit law-assured investors while only providing barely direct benefits for village citizens. Reflecting the controversy above, the capital-directed *BUMDes* that received its status as a legal entity will only experience further corporatization that holds no substantive meaning towards the realization of village prosperity.

Cooperatives present an alternative and healthy model for enterprise as they further contribute towards sustainable development well beyond job creation. The inclusion of person per person in Cooperatives enables them to develop well beyond the framework of regular business entities and their involvement in the triple-co principles of co-ownership, co-determination, and co-responsibility will result in a gradual but willing entrepreneurship of the village community that can provide employments not only in cities but also in villages. The regretful notion of *BUMDes* as a platform of capital accumulation alone with no place for the entire village community to participate in may prevent *BUMDes* from really contributing to social prosperity as cooperatives potentially can. This presents an underlying problem of *BUMDes* being its structure that ignores the important point that individuals can really play in a framework of voluntary economic-social institutions that is well reflected in cooperatives.

As a result, cooperatives appear as the original alternative which fully reflects Indonesia's traditional values of cooperativism, togetherness, and joint effort based on kinship. *BUMDes* either can incorporate the framework of Cooperatives in its regime or it can admit its faults and champion cooperatives as the true economic institution of Indonesia's economy. As a further reflection, cooperatives can even fit in the context of the United Nations Sustainable Development Goals (SDGs) that Indonesia currently championed as it can be aimed to end poverty and hunger and also to promote inclusive and sustainable economic growth for all people. Research shows that in realizing these objectives, cooperatives have several engagement strategies and benefits under its belt which include the promotion of access to markets; the realization of a higher purchasing power; the minimization of cost; the advancement of skill and knowledge; and the improvement of livelihoods. Moreover, investment is also included as an engagement strategy, yet it is not a priority and should be followed by real socio-economic development. The main purpose of investment strategy is also manifested through affordable financial services to small producers or individuals. (*Cooperatives and the Sustainable Development Goals* 2017). Concluding from the above explanations, cooperatives have different focuses compared to the worldwide economic system which is dominated by corporations. It focuses on members and their local needs and plays a true strategic role in economics. If it is taken seriously, it can be the central agent to the realization of sustainable and people-centered economic development around the world.

5 Conclusions

It is indeed obvious that the revitalization of *BUMDes* in the omnibus in lieu law is an instance of an economic shock in Indonesia's economy as it portrays a sudden shift from *BUMDes* as a business entity localized in a village into a legal entity of capital accumulation that can be considered under constant threat of further corporatization by recent implementing regulations. Furthermore, BUMDes regulation can be considered as a preposterous legal framework as instead of reinvigorating cooperatives which has been mandated by Indonesia's constitution, the government chose to suddenly reinvigorate *BUMDes* as a new legal entity which have fundamental issues in its legislation and potentially problematic in its implementation.

The omnibus in lieu law along with its main motive in bringing capital as a central theme became the very background of the BUMDes regulations. From the existing BUMDes legislation, it is clear that BUMDes has been directed to become a similar entity to a limited company. In its execution, *BUMDes* can encourage an increase in the village's income but such an income cannot be felt directly by the village community. Therefore, this research recommends that *BUMDes* reflects on cooperatives both as values and as an institutional idea to determine its relevance in Indonesia's economy. One important consideration is that the form of a limited company will truly harm *BUMDes*' independence since it is based on capital ownership and not on shared ownership of villagers. Thus, the most capitalized will determine the course of *BUMDes*.

Instead, the character of Indonesia's economic democracy should always become the key point to regulate and develop *BUMDes*. It must prioritize rural community involvement and socio-economic development. Thus, it can be concluded that Cooperatives is the most suitable institutional form and framework within which *BUMDes* can develop. Cooperative as the association of people has a resolute orientation towards community empowerment and it can be manifested as the triple-co principles which will enable the rural community to co-owned the shares; co-determine the enterprise actions; and be co-responsible for the development of *BUMDes*. In its aftermath, healthy *BUMDes* can be reliable in supporting Indonesia's economy according to its values and ideals.

References

Amanda HW (2017) Strategi Pembangunan Desa dalam Meningkatkan Pendapatan Asli Desa Melalui Badan Usaha Milik Desa (BUMDes) (Studi Pada Badan Pengelolaan Air Minum (BPAM) di Desa Ketapanrame Kecamatan Trawas Kabupaten Mojokerto). Jurnal Fakultas Ilmu Sosial 4(1):7–12

Clemente-Suárez VJ, Navarro-Jiménez E, Moreno-Luna L, Saavedra-Serrano MC, Jimenez M, Simón JA, Tornero-Aguilera JF (2021) The impact of the covid-19 pandemic on social, health, and economy. Sustainability 13(11):1–25. https://doi.org/10.3390/su13116314

Dirkareshza R, Sihombing ENAM (n.d.) Acceleration of village welfare through Bumdes: disorientation of implementation of Bumdes regulations and policies, 8–11. https://doi.org/10.30641/dejure.2021.V21.419-434

Dolfriandra Huruta A, Kundhani Y, Manurung CR (n.d.) The development of village-owned enterprises: lessons learned from Ponggok Village Pengembangan Badan Usaha Milik Desa (BUMDes): Belajar dari Desa Ponggok

Eriksson L (2015) Social norms theory and development economics background paper to the 2015 world development report. http://econ.worldbank.org

Herawati E (2016) Badan Usaha Milik Desa, Status dan Pembentukannya. Bina Nusantara Online

Hatta M (1954) Membangun Koperasi dan Koperasi Membangun dalam Meninjau Masalah Koperasi. PT. Pembangunan Djakarta, Jakarta, p 41

Hatta M (1971) Cita-Cita Koperasi Dalam Pasal 33 UUD 1945 dalam Kumpulan Karangan Dr. Mohammad Hatta, Jakarta: Pusat Kooperasi Pegawai Negeri Djakarta-Raja, pp 331–337

Hatta M (1975) "Menuju Negara Hukum", speech at the conferment of his honorary doctorate degree from Universitas Indonesia, 30 Agustus 1975, Jakarta

ILO (2015) Cooperatives and the sustainable development goals: a contribution to the post-2015 development debate. www.ica.coop
J Econ Hist 59(4):888–889
Kahraman C, Orobello C, Cirella GT (2022) Changing dynamics with COVID-19: future outlook. In: Cirella GT (ed) Human settlements: urbanization, smart sector development, and future outlook. Springer, Singapore, pp 235–252
McGee R (2022) The governance shock doctrine: civic space in the pandemic. Dev Policy Rev. https://doi.org/10.1111/dpr.12678,pp.1-22
Meliala A, Melawan Dinamika Kapitalisme Dengan Hukum Ekonomi Kerakyatan. PT. Rajagrafindo Persada, Depok, p 70
Pemerintah targetkan jumlah BUMDes berkembang capai 10.000 pada 2024 - ANTARA News (n.d.). https://www.antaranews.com/berita/2049998/pemerintah-targetkan-jumlah-bumdes-berkembang-capai-10000-pada-2024
Pengamat: Dana Desa Rp 1 Miliar Jadi "Jualan Politik" (n.d.). https://nasional.kompas.com/read/2014/03/12/1455139/Pengamat.Dana.Desa.Rp.1.Miliar.Jadi.Jualan.Politik
Pulungan MS (2021) Familyhood principle and the cooperatives ideas in economic provisions in the Indonesian Constitution - Dialnet. https://dialnet.unirioja.es/servlet/articulo?codigo=7746806
Simamora EE, Ester J, Kristy DW (2022) Analisis Status Badan Hukum Badan Usaha Milik Desa Dengan Berlakunya Undang-Undang Nomor 11 Tahun 2020 Tentang Cipta Kerja dan Peraturan Pelaksananya. Jurnal Hukum Bisnis Bonum Commune 5(2):161–169
Sumiasih K (2019) Peran BUMDes Dalam Pengelolaan Sektor Pariwisata (Studi di Desa Pakse Bali, Kabupaten Klungkung). Jurnal Magister Hukum Udayana 7(4):565–585
Swasono S-E (2018) Keindonesiaan: Demokrasi Ekonomi, Keberdaulatan, Kemandirian. USTPRESS, Yogyakarta, pp 14-15
Utami KS, Tripalupi LE, Meitriana MA (2019) Peran Badan Usaha Milik Desa (Bumdes) dalam Peningkatan Kesejahteraan Anggota Ditinjau Melalui Kewirausahaan Sosial. Jurnal Pendidikan Ekonomi 11(2):498–508
UU Perkoperasian Dibatalkan Karena Berjiwa Korporasi (n.d.). https://www.hukumonline.com/berita/a/uu-perkoperasian-dibatalkan-karena-berjiwa-korporasi-lt5385bfa83b01f

Environmental Footprint for the Accumulation of Plastic Waste During the COVID-19 Pandemic

Cengiz Kahraman, Mohammed Sharaf, and Giuseppe T. Cirella

Abstract This chapter examines the environmental consequences of excessive usage and improper disposal of plastics during the COVID-19 pandemic. The significant ecological impacts of plastic pollution during this period resulted from the high demand for single-use plastics and personal protective equipment (PPE), the increased production of medical waste, and the prioritization of incineration and landfilling. With the commencement of vaccination campaigns, the issue of generating plastic-based medical and household waste has further intensified. It is recommended that appropriate strategies for managing plastic waste, encompassing comprehensive and step-by-step approaches that involve adequate sorting, sanitization activities, proper technological intervention, and process optimization be implemented. In conclusion, pivotal solutions and recommendations such as the adoption of bioplastics and the substitution of fossil-fuel-based plastics with biodegradable or compostable materials could be utilized in future scenarios.

Keywords Sustainability · Single-use plastics · Personal protective equipment · Vaccine · Household and medical plastics waste · COVID-19 crisis

1 Introduction

To mitigate the spread of the COVID-19 pandemic, stringent lockdown measures were implemented worldwide (Kahraman et al. 2022). As a predictable outcome, the global demand for fossil fuels experienced a sharp decline, leading to a substantial decrease in oil prices. Consequently, the cost of producing plastics derived from fossil fuels became more economical in comparison to recycling practices (Kimini 2020;

C. Kahraman (✉) · M. Sharaf
Department of Engineering Sciences and Department of Maritime Transportation Management Engineering, Istanbul University-Cerrahpaşa, Istanbul, Turkey
e-mail: cengiz.kahraman@istanbul.edu.tr

G. T. Cirella
Faculty of Economics, University of Economics, Sopot, Poland
e-mail: gt.cirella@ug.edu.pl

G. T. Cirella and B. Dahiya (eds.), *City Responses to Disruptions in 2020*, Advances in 21st Century Human Settlements, https://doi.org/10.1007/978-981-99-7988-2_7

Adyel 2020). This cost incentive, coupled with lifestyle changes that necessitated increased plastic consumption, posed a significant challenge in addressing the issue of plastic pollution.

The COVID-19 pandemic has witnessed a surge in the utilization of single-use plastic products such as face masks, gloves, and specific packaging materials. This surge has resulted in elevated levels of emissions and other pollutants. These materials have littered both terrestrial and marine environments, causing undeniable harm to ecosystems and wildlife. In light of the pandemic, there is a pressing need for a global reassessment of single-use plastic production, consumption, and waste management practices. This reevaluation is crucial to equip societies in managing the ongoing impacts, as well as to prepare for potential future crises or similar disruptive events.

Moreover, plastic plays a pivotal role in safeguarding individuals, particularly frontline healthcare workers such as doctors, nurses, and other medical personnel during pandemic-oriented conditions. As a fundamental preventive measure, it is mandated for healthcare professionals to wear personal protective equipment (PPE) when dealing with suspected cases.

The healthcare system faced immense challenges during the COVID-19 pandemic, exposing its vulnerabilities to severe stresses. The capacity of hospital beds, availability of medical supplies, and access to PPE were severely impacted, highlighting the lack of preparedness for such an unprecedented crisis. The situation in North Italy serves as a well-known example, where the healthcare system teetered on the brink of collapse (McCarthy 2020). As such, hospitals are significant consumers of energy through electricity usage, transportation, and the manufacturing, utilization, and disposal of healthcare products. As a result, the healthcare sector is responsible for approximately 4.4% of global net emissions, making it a substantial contributor to the world's environmental footprint (Bawaneh et al. 2019; Karliner et al. 2019, 2020). Figure 1 illustrates that space heating, ventilation, water heating, and cooling constitute the primary energy-consuming components, accounting for 63% of the total energy consumption worldwide. The rigorous standards for air quality significantly contribute to the substantial energy usage associated with heating and cooling operations. Ventilation, cooling, and lighting systems demand significant amounts of energy. Notably, space heating, especially in inpatient centers, heavily relies on natural gas as the primary energy source.

The potential rise in healthcare-related environmental footprints is primarily driven by increased energy requirements due to stringent hygiene measures. The heightened demand for food packaging and disinfectants contributes to the expansion of these footprints (Gallego-Schmid et al. 2019). Notably, a significant factor is the extensive testing needed, encompassing viral and antibody tests (Walensky and Del Rio 2020). Testing activities generate environmental footprints primarily through transportation processes (FDA 2020). It is also important to acknowledge that the environmental footprints resulting from transportation can vary depending on energy consumption and specific circumstances (Lispasto 2017). In 2020, the conduction of testing, including the use of swabs, plastic bottles, vials, and chemical reagents, is associated with environmental footprints and resource consumption. Proper waste treatment and disposal methods are essential for minimizing these footprints. Table 1

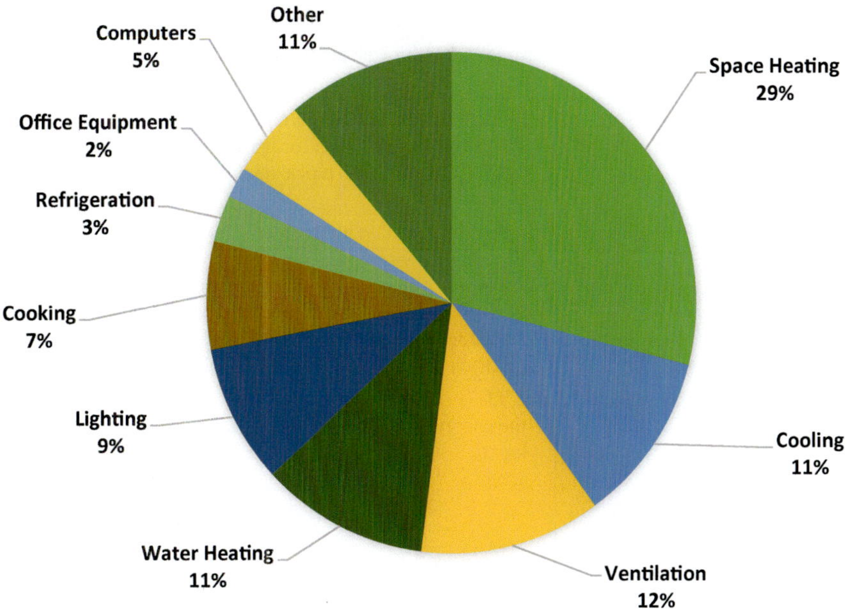

Fig. 1 Major fuel consumption breakdown by end-use throughout the world in 2020. *Source* adapted from Bawaneh et al. (2019) and Karliner et al. (2020)

presents the associated energy consumption and estimated demand for PPE and test kits, further highlighting the environmental considerations involved.

Plastics serve as the predominant material used in the manufacturing of PPE gear (Klemeš et al. 2021). A portion of the energy consumed in this process can potentially be recovered, thereby offering an avenue for mitigating its associated footprint. This presumption, however, requires thorough evaluation and management of solid waste to determine its feasibility and quantifiable impact in the aftermath of the COVID-19 pandemic, i.e., a process that cannot be accurately assessed in the short term.

2 Environmental Footprint and the Filtration Efficiency of Masks

In the midst of the crisis, face masks have emerged as a high-demand PPE item. In the context of everyday mobility, the utilization of face masks extended beyond healthcare facilities and was mandated for the general public in many countries around the world. However, a severe shortage in the supply of face masks was initially experienced, with the demand during the pandemic surpassing immediate availability. The selection of masks is guided by various criteria, including filtration efficiency (for particle size range), material composition, fit (with or without gaps), ease of

Table 1 PPE demand and energy consumption estimated in 2020

PPE devices and weight	Estimated demand	Estimated energy consumption and explanatory notes
Mask, 12–13 g/pcs	129 G/month	1.29 TWh/month = ~4.6 PJ/month – the energy consumption for masks production is 0.000792–0.0342 kWh/pcs, taking 0.01 kWh/pcs for the estimation
Gloves, ~7 g/pair	65 G/month	65 G/month 1.95 GWh/month = ~7.02 TJ/month – the energy consumption for glove production is 3×10^{-5} kWh/pair, plus electricity consumption in each plant is 41.19–77.84 kWh/tons for concentrated latex where the centrifuge process required the greatest amount of electricity accounting for 94–97% of total consumption
Fast testing kit (without gloves), ~14 g/test	390 M – from January 2020 to August 2020	168 TJ – the energy consumption of plastics production is 30.82 MJ/kg
Shield, ~ 81 g/pcs	1.59 M/month	3.9 TJ/month – the energy consumption of plastics production is 30.82 MJ/kg

Source Ecochain (2020), Prata et al. (2020), Allison et al. (2020), Klemeš et al. (2020a, b), and Konda et al. (2020)

handling, and breathability. Availability and cost also influence the decision-making process.

The environmental impact associated with face masks encompasses various stages, including production, packaging, transportation, washing and disinfection practices, end-of-life treatment, and ultimate disposal. This impact involves factors such as energy consumption, pollutants, and other related considerations. The overall environmental footprint of face masks can be significantly influenced by variables like washing and disinfection methods, frequency of usage, and approaches to end-of-life treatment. In order to curb the transmission of respiratory illnesses, physical interventions have been deemed necessary. Randomized control tests (RCTs) and observational studies were conducted to evaluate the effectiveness of these interventions (Das et al. 2021; Tripathi et al. 2020; Data OWi 2022; Phadke et al. 2021; Zehrung et al. 2017). Overall, face masks emerged as one of the standardized government intervention approaches for the general population.

The primary criterion for selecting a mask is its filtration efficiency, typically expressed as a percentage. However, a high filtration percentage does not guarantee complete protection. The particle size used for measuring filtration efficiency is typically 300 nm, which is the standard size applied in certification tests for masks (Table 1). The main constituent of an N95 mask, for instance, is polypropylene. When transportation is not taken into account, the embodied pollutants of an N95 mask is estimated as 0.05 kg CO_2 eq/single-use (Ecochain 2020). One could notice that transportation contributes to the high share of emissions of surgical masks (Allison et al. 2020). Moreover, it shows the high share of emissions in terms of mask cleaning. As such, it is possible to reduce the increase in waste products and emissions through strategies such as increased reuse of masks. For instance, if a mask is reused 183 times, the equivalent emission could be lowered to 0.036 kg CO_2 eq. Evaluating the filtering efficiency in relation to the emissions invested requires considering the effectiveness and environmental footprint of advanced reusable masks. Based on available data, cloth masks exhibit a reasonable performance index compared to surgical masks. However, assessing the user phase is complex and requires further consideration, as it is influenced by individual behaviors that can vary from person to person. In the case of single-use disposable masks, proper waste treatment is necessary to minimize impacts through energy recovery methods.

Germany provides another prominent illustration of the environmental footprint associated with masks. As a large country, Germany employs over 800,000 individuals in the hospital sector, as reported by the bureau of statistics. On a monthly basis, the country consumes approximately 17 million FFP masks, resulting in a CO_2 emission rate of 850 tons for N95 protective masks alone (Ecochain 2020; Bown 2020). To put this into perspective, the emissions from these masks would be equivalent to driving a distance of 42.5 million kilometers, which is equivalent to circling the globe 1,060 times. The German minister of environment claimed that in 2020 Germany might need up to 12 *billion* masks with an environmental footprint of around 720,000 tons of CO_2-equivalent (Ecochain 2020). Consequently, this puts one person's annual CO_2 budget at 2.3 tons if we want to fight climate change effectively. This means that the need for protective masks in Germany alone used up the CO_2 budget for more than 310,000 people. A tiny change in a product can make a big difference on a large scale, e.g., by a simple change of material with face masks, we could reduce their impact by 20%, saving 144,000 tons of CO_2 emissions (Ecochain 2020).

3 The Environmental Footprint from the Launching of the Vaccination Drive

The beginning of 2021 marked the commencement of COVID-19 vaccination campaigns on a global scale. Governments around the world were making concerted efforts to vaccinate as many individuals as possible, aiming to mitigate the impact of the COVID-19 pandemic. In order to meet the global demand for widespread immunization, vaccine production was accelerated. However, this accelerated production resulted in a significant increase in plastic-based medical and household waste. With the initiation of vaccination drives, this waste issue was expected to further escalate. As a consequence, the disposal of vaccine containers led to the generation of approximately 1×109 tons of plastics, rubbers, and solid wastes (Das et al. 2021). Disposables used in the vaccination process include syringes, droppers, vaccine bottle caps, and gloves. For instance, considering an average syringe weight of approximately

5 g and a population of around 7 × 109 individuals, the use of syringes alone results in the generation of approximately 1 × 109 tons of single-use plastics (SUPs), not accounting for other peripheral plastic materials such as gloves and masks. Another example from the United States, where inoculating the entire population would require approximately 650 × 106 vaccine doses, leading to a substantial amount of waste (Data OWi 2022). Disposables, including syringes, droppers, vaccine vials, bottle caps, gloves, and portable storage containers, are necessary components of the vaccination process, with specific temperature requirements. Consequently, the management of the biomedical waste generated during vaccination campaigns needs to be addressed with careful consideration. Therefore, the accumulation of SUPs waste during inoculation campaigns presents a significant environmental footprint, particularly in developing and low-income countries.

3.1 Exploring Alternative Vaccine Packaging Materials and Delivery Options

In this subsection, we will focus on a select few innovations that are relevant. The COVID-19 pandemic has highlighted the need for effective packaging materials that can ensure vaccine efficacy (Phadke et al. 2021; Zehrung et al. 2017). Customized storage conditions during transportation are essential for global distribution, requiring the use of various packaging products such as bags, vials, ampoules, special freezers, and containers (Das et al. 2021; Phadke et al. 2021; Zehrung et al. 2017; IMD 2020). These products must meet high-performance standards to ensure proper handling. Traditional borosilicate glass containers have drawbacks, including a tendency to break and potential delamination, which may impact drug efficacy (IMD 2020). To address these concerns, alternative materials such as aluminosilicate glass are being explored to reduce contamination from broken particles and minimize crack formation (Das et al. 2021; Phadke et al. 2021; Zehrung et al. 2017). Innovations such as phase-changing and smart materials are also being investigated to create suitable freezer containers that can meet the ultra-cold temperature requirements for vaccine storage and transportation (Zehrung et al. 2017). Moreover, the use of smart packaging can enhance supply chain transparency and security.

Honeywell introduced Aclar EdgeTM polymer bottles and vials for pharmaceutical packaging (FluoroCarbon 2021). These containers are made from poly-chloro-tri-fluoro-ethylene (PCTFE) and offer several advantages. They are lightweight, reducing transportation costs by 75% compared to their glass counterparts. The company claims that these containers eliminate the hazards associated with breakage (FluoroCarbon 2021). Traditional syringes are made of recyclable polypropylene, with a shortcoming that only ~1% is recycled. Currently, the idea of prefilled syringes is being used. These syringes are composed of cycloolefin polymers and copolymers plastic materials. It is highly recommended that prefilled syringes be adopted. Unlike overfilled vials, prefilled syringes minimize wastage, ~20–30% (WHO 2021).

The packaging waste is reduced by ~30% when secondary packaging of prefilled syringes is applied and the overall waste by ~50% (WHO 2021). Cyclic olefin polymers and copolymers when compared with PP have lower oxygen and water vapor permeability, thus allowing for long-term storage. Also, they are compatible with sterilization processes due to their heat-resistant nature (Christopher Weikart 2020).

The vaccine packaging domain needs to be influenced by sustainability as a major trend. For the reduction of packaging waste, scientists are looking into compostable and biodegradable secondary packaging options (Christopher Weikart 2020; Phadke et al. 2021; Zehrung et al. 2017). This includes shipping coolers derived from plant or recyclable content. The vaccine packaging and value chain community are manipulating several routes for the development of materials for primary, secondary, and tertiary packaging that will ensure the efficacy and effectiveness of vaccines and other pharmaceuticals, in general.

3.2 Environmental Footprints of Household Uses of SUPs as Domestic Precautionary Measures Against COVID-19

Social distancing and lockdown measures were implemented to mitigate the spread of COVID-19. As such, the worldwide closure of restaurants, cafeterias, and other recreational establishments had resulted in a widespread reliance on home delivery services. Accordingly, there had been a significant rise in the production of plastic packaging materials to meet the escalating demand. Various forms of plastic packaging have experienced a substantial increase during the first year of the pandemic (Ncube et al. 2020). Notably, South Korea witnessed a staggering 92.5% surge in online grocery shopping during the pandemic compared to 2019 (Mu-Hyun 2020). Similarly, numerous countries across different continents experienced a significant rise in online shopping, with the average increase ranging from 12 to 57% (Mu-Hyun 2020). In the United States, online shopping surged by 50% during the lockdown (Rattner 2020). The prevalent packaging materials used for grocery items typically include films, foams, and multilayer plastics (Ncube et al. 2020, UNEP 2018).

In 2020, the production of SUPs accounts for 46% of the global plastic waste. China stands as the primary global producer and generator of SUP waste, followed by the United States, which holds the highest per capita generation of SUP waste, Japan, and the European Union (Das et al. 2021; Tripathi et al. 2020). Despite the lockdown measures coming to an end, the demand for SUPs is expected to continue rising. The COVID-19 pandemic has led to the withdrawal of policies aimed at reducing SUP usage, particularly with regard to plastic bags. However, many consumers who have already embraced non-plastic alternatives voiced their objections to this withdrawal (Xanthos and Walker 2017). Claims suggesting that reusable bags contribute to the transmission of the disease are highly questionable when considering proper hygiene practices. As a result, the utilization of PPE has resulted in a significant surge in plastic pollution. The pandemic led to a substantial increase in the daily production of single-use face masks, reaching a staggering 116 million, which is twelve times the usual quantity (Bermingham and Tan 2020). To address the heightened demand, the World Health Organization issued a recommendation to increase the production of disposable PPE gear by 40% (WHO 2020).

Following the conclusion of the lockdown, in 2021 and 2022 it was anticipated that individuals would limit their use of disposable face masks. Consequently, an estimated 129 billion masks and 5 billion gloves were utilized each month (Prata et al. 2020). Illustratively during the year following the lockdown in the United States, it alone could produce an entire year's worth of medical waste within a span of two months (Bengali 2020). In Wuhan, China, the epicenter of the outbreak, hospitals generated 24 tons of face masks, gloves, and personal protective gowns in 2021, which amounted to six times the daily average prior to the pandemic (Cutler 2020). Amidst the lockdown, there was a notable increase in the demand for plastic driven by individual preferences. A concrete example was observed in Singapore, where the 8-week lockdown resulted in an additional 1400 tons of plastic waste generated from the consumption of take-out meals and home-delivered groceries (Bengali 2020). Moreover, the global plastic packaging market unexpectedly rose to a staggering valuation of 1012.6 billion in 2021, showcasing a compound annual growth rate of 5.5%. This growth can be largely attributed to the response of governments during the pandemic (Markets 2021).

As such, the COVID-19 pandemic exerted significant pressure on conventional waste management practices, leading to the adoption of mismanaged strategies such as mobile incineration, direct landfills, and local burnings (Silva et al. 2020). For example, the improper disposal of a mere 1% of face masks resulted in an astonishing quantity of over 10 million items, weighing approximately 40 metric tons (WWF 2020). Furthermore, the presence of discarded and waterlogged plastic in coastal areas, directly associated with COVID-19, sparked concerns regarding the heightened potential for microplastic pollution (Stokes 2020). While the advantages of plastics, including their affordability, lightweight nature, and ease of processing, are widely acknowledged, the exponential growth in plastic production witnessed over the past century has presented significant challenges. Unless we engage in fresh thinking and reevaluate how we manufacture, utilize, and manage plastics, we will struggle to effectively address the vast quantities of plastic waste generated. Paradoxically, despite the recognized benefits, plastic is now regarded as a major environmental threat and a significant contributor to pollution due to its rampant littering and mismanagement as solid waste. The COVID-19 pandemic impacted our health and our economies, and it had disruptive effects on our personal lives. The reaction to the first wave of COVID-19, for instance, had an impact on the SUPs, as follows:

- The noticeable rise in littering of PPE, particularly SUPs such as face masks, gloves, and specific types of packaging. This trend resulted in additional emissions and other pollutants during the production process and subsequent open incineration, highlighting the importance of implementing proper waste management practices.

- During the pandemic, there was a global surge in both imports and domestic production of face masks and gloves to meet the increased demand.
- The significant increase in the consumption of takeaway and home delivery food during lockdowns led to a corresponding surge in the usage of SUPs food containers. This has contributed to the overall volume of SUPs being utilized.

The prioritization of human health over environmental health further resulted in a delay, and in some cases, a reversal of policies aimed at reducing the usage of SUPs (WWF 2020). Throughout the pandemic, recycling efforts slowed down, and the demand for recycled plastic materials declined due to reduced profitability in recycling operations. Consequently, there was a noticeable increase in the environmental impact associated with plastics (WWF 2020). Undoubtedly, it was imperative to embrace circular economy approaches that prioritized recycling practices and enforce stringent policies against plastic pollution. As such, companies, in particular, have a responsibility to minimize their reliance on virgin plastic and instead increase the utilization of recycled plastic materials to fulfill their corporate social and environmental obligations. To achieve the United Nations Sustainable Development Goals adopted by the Member States of the United Nations, it is crucial to make a collective and dedicated effort to protect the environment during and beyond future crises and shock events. Embracing the sustainability concept, which underscores the importance of preserving our natural resources and fostering ecological harmony, becomes imperative in this pursuit (Cirella et al. 2007; Xiao and Torok 2020). This commitment not only aligns with our responsibility to current and future generations but also reinforces the fundamental principles of sustainable development (Cirella and Zerbe 2014; Cirella 2020, 2022; Cirella et al. 2020, 2021; Fu et al. 2022; Jiao et al. 2021; Kahraman et al. 2022; Lewandowski and Cirella 2022; Oniszczuk-Jastrząbek et al. 2020; Spring and Cirella 2022).

4 Plastic Policies Under the COVID-19 Pandemic and Potential Environmental Consequences

Significant advancements have been observed in plastic sustainability and solid waste management in recent times. However, the emergence of the COVID-19 pandemic has brought about considerable challenges in the realm of plastic usage and management. Various precautionary measures, such as partial or complete lockdowns of cities and regions, were implemented (Tobías 2020; Xiao and Torok 2020). As a consequence, there has been a widespread surge in the utilization of SUPs-based PPE by the general public, including masks and gloves (Tobías 2020; Xiao and Torok 2020). Due to concerns about hygiene and the implementation of preventive measures, consumer behavior underwent a significant shift (Jribi et al. 2020). Moreover, the plastic industry highlighted the potential risks of cross-contamination associated with reusable containers and bags (Schnurr et al. 2018). This led to actions taken by various regions, such as the province of Newfoundland and Labrador in Canada and the states of New York, Delaware, Maine, Oregon, Connecticut, and Hawaii in the United States, among others, to raise awareness and reintroduce SUPs products while even banning the use of reusable alternatives. Similarly, the implementation of deposit return schemes for soft plastics had been postponed. Additionally, there was an increased demand for plastic suppliers, particularly in China and the United States, due to the heightened use and consumption of SUPs during the pandemic (Bown 2020). The decrease in water, land, and air transport also resulted in a significant drop in oil prices, favoring the production of virgin plastics rather than recycled plastics.

The widespread adoption of PPE by the general public, extended beyond frontline healthcare workers, witnessed a substantial increase. Consequently, there was a surge in the production of SUPs to meet the heightened demand. Notably, China experienced a remarkable 450% increase in face mask production, escalating from 20 to 110 million masks per month (Ecochain 2020; Bown 2020). The market for N95 respirators also saw a significant rise, with production surging from approximately 200,000 to 1.6 million (Ecochain 2020; Bown 2020). However, controversy arose surrounding the usage of masks by ordinary citizens due to improper handling and disposal practices, compounded by shortages in healthcare facilities (Bermingham and Tan 2020). To prevent contamination, surgical masks should have been disposed of properly and not worn for more than three hours. Unfortunately, mismanaged disposal of PPE rapidly spread in various public spaces and natural environments (Prata et al. 2020). For example, during an environmental survey conducted on a Hong Kong beach, numerous discarded masks were found to be littered around the area (Oceans Asia 2020). Masks are typically made with nonwoven materials that often contain polypropylene and polyethylene, which may break down into smaller microplastic particles (Prata et al. 2020). Highlighting the hazards of plastic pollution, it has been discovered that microplastic microfibers found in both water and sediment of the Magdalena River in Colombia were attributed to the degradation of nonwoven fiber textiles (Martínez Silva and Nanny 2020).

Undoubtedly, the significant increase in the utilization and disposal of masks, gloves, plastic bags, and other SUPs, particularly in certain business sectors during the post-pandemic era is concerning. For instance, protective plastic films are employed in chairs and cash machines to mitigate the risk of contamination through air droplets. This surge in SUPs and PPE waste has created a substantial challenge in implementing proper waste management practices (Prata et al. 2020). Many countries have classified all medical and domestic waste associated with COVID-19 as contagious waste (EC 2020). The ability to address this crisis varies across different countries, particularly along the divide between rich and poor nations. Wealthier countries have the resources to manage such waste effectively through high-temperature incineration followed by proper landfilling of residual ash (Windfeld and Brooks 2015). However, in countries with fewer resources, improper waste management practices such as open burning and direct landfilling are often resorted to. This is particularly prevalent in the southern hemisphere, where numerous municipalities lack the necessary resources to handle the mounting quantities of waste being generated (Hoornweg and Perinaz 2012). These countries face significant logistical challenges in waste management provisions. In an effort to minimize the risk of pollution, the incidence of local burning and uncontrolled landfilling of plastic waste has increased considerably (Corburn et al. 2020). In contrast, wealthier countries have taken appropriate measures to manage plastic waste in the context of combating COVID-19. For example, in Wuhan, China, mobile incinerators were utilized to handle a four-fold increase in infectious waste during the pandemic (Saadat et al. 2020).

The COVID-19 pandemic, albeit for a brief period, had a positive impact on certain aspects of human life. During this time, there was an observable improvement in air quality, a decrease in smoking rates, a reduction in household food waste, and a significant decrease in emission levels. However, the long-term consequences on the environment are concerning due to the substantial increase in the usage and improper management of SUPs and PPE (Prata et al. 2020). This has resulted in an expanded environmental footprint caused by the incineration, leakage, and burial of a significant amount of plastic waste, particularly SUPs and PPE. Consequently, considerable energy embodied in these plastic materials has been lost, leading to substantial releases of emitted pollutants and hazardous compounds into the environment, thereby exacerbating the environmental footprint (Heidari et al. 2019). Furthermore, lightweight items such as face masks, gloves, and other PPE and SUPs pose a threat to aquatic and terrestrial ecosystems as they can easily be carried by the wind and deposited in surrounding natural habitats. Just a 1% improper disposal rate of masks alone can introduce over 10 million masks into the environment on a monthly basis (WWF Italy 2020). Considering an average weight of approximately 3–4 g per mask, this would equate to approximately 30,000 to 40,000 kg of masks finding their way into natural environments (WWF Italy 2020; Fadare and Okoffo 2020). Consequently, the degradation of such plastic items in the environment would result in the formation of microplastics. Thus, the COVID-19 pandemic placed a heavy burden on plastic waste management, the environment, and human health, and has had additional repercussions on the global economy and societal systems.

The crisis of 2020 presented numerous challenges while also offering an opportunity to implement effective solutions to combat plastic pollution. To ensure the proper management of infectious plastic-based waste generated during the pandemic, it was crucial to address various procedural hurdles, including source identification and segregation processes. As a response, several international organizations such as UNICEF, WHO, European Commission, Asian Development Bank, UN-Habitat, and United Nations Environment Programme, along with national health organizations and prominent pollution monitoring bodies such as the Ministry of Ecology and Environment in China, Occupational Safety and Health Administration in the USA, Central Pollution Control Board in India, Ministry of Health and Indigenous Medical Services in Sri Lanka, National System for Environmental Protection in Italy, and Ministry of Health and Population in Nepal, have recently released comprehensive guidelines and advisories. These guidelines aim to provide clear instructions for the proper handling of biomedical and household waste. It is essential that these guidelines are implemented both at the international and national levels, as summarized in Appendix 1, Table 2.

5 Policy Recommendations

The COVID-19 pandemic has highlighted the reliance on disposables, including SUPs and PPE, and has revealed the vulnerabilities in waste management and plastic reduction policies. It is crucial to prioritize strategies that reduce plastic usage while safeguarding both health and the environment. A key consideration should be the adoption of sustainable and comprehensive solutions that encompass the entire life cycle of plastics (Silva et al. 2020). The significant environmental impact of plastic pollution during the pandemic can be attributed to several factors: (1) increased demand for SUPs, (2) heightened demand for PPE, (3) a surge in medical waste generation, and (4) a focus on incineration and landfilling as disposal methods. Based on these observations, the following three recommendations are proposed.

First and foremost, a crucial step in addressing plastic pollution is the redesign of plastics and a shift away from hydrocarbon-based resources. This transition aligns with international agreements aimed at achieving a green and circular economy. However, it requires a supportive legislative framework and a clear commitment. One potential solution lies in the use of bio-based plastics, which offer a sustainable alternative. Currently, bio-based plastics account for less than 2% of the total 348 million tons of plastics produced, primarily due to the lower cost of fossil-based plastics (Imbert et al. 2019; Carus 2018; Hatti-Kaul et al. 2020). Several factors contribute to this limited market share, including extensive land use, financial investments, and the lack of established recycling and disposal options (Chanprateep 2010). To overcome these challenges, the concept of biorefinery, which involves synthesizing raw materials from biomass by-products and waste streams, emerges as a viable alternative (Hatti-Kaul et al. 2020). This approach has the potential to yield high-performance bio-based polymers with desirable material properties (Chanprateep 2010; Zhu et al. 2016). Such polymers can be utilized in the production of various items, including face shields, respirators, ventilators, space/counter dividers, syringes, and boots (Hatti-Kaul et al. 2020; Ioannidou et al. 2020; Chanprateep 2010; Zhu et al. 2016). With proper decontamination, these items can be reused and eventually undergo energy recovery. Furthermore, the development of bio-based solutions for food packaging, masks, gloves, and other disposable plastics can significantly reduce their environmental footprints. However, a dynamic assessment of the life cycles of these bio-based polymers is essential. A comprehensive comparison should be made in terms of sustainability, considering both the environmental footprint of plastic production waste and alternatives such as paper bags (Klemeš et al. 2020a, b). The shift from fossil-based plastics to biobased plastics should be accompanied by the development of an enhanced circular and integrated system. While the COVID-19 pandemic may have prompted significant changes in economic structures, it is crucial to remain committed to sustainable plastic solutions and not deviate from such a trajectory.

Another important aspect to address is the reduction of plastic waste by minimizing the use of SUPs and PPE. It is advisable to utilize either plastic or reusable fabric bags for packaging purposes. Additionally, promoting the use of reusable PPE instead of single-use alternatives, along with proper hygiene practices and sterilization, can help alleviate concerns regarding reusable packaging and PPE (Chin et al. 2020). Taxes and bans on fossil-fuel-based SUPs should be maintained, as exemplified by the notable reduction of 90% in SUP bag usage following the introduction of a small fee of 15 cents per bag in Ireland in 2002 (Xanthos and Walker 2013). Consequently, public awareness of plastic pollution must be heightened through social media, government initiatives, and stakeholder involvement (Schnurr et al. 2018). Similarly, solutions for reusable PPE should be implemented during pandemics. In order to establish effective waste management practices during the pandemic, a sustainable policy should incorporate the 10 Rs: refuse–reject, rethink, reduce, reuse, repair, refurbish, remanufacture, repurpose, recycle, and recover (Potting et al. 2017). It is important to emphasize that raising public awareness about the adverse effects of plastic pollution is crucial in mitigating plastic leakage. By exerting pressure on governments and industries, this type of change can drive them to adopt similar practices. Evaluating recycling, composting, and waste-to-energy incineration as end-of-life solutions for plastics should also be considered.

Lastly, it is crucial to optimize the management of plastic waste. The COVID-19 pandemic resulted in a significant global increase in medical waste and SUPs. These circumstances necessitate stronger regulations for the tracking, treatment, and disposal of medical waste, not only at the national level but especially at the international level, to facilitate the development of strategies and enhance efficiency in medical waste management. The optimization of pre-sorting and disinfection processes can improve the potential recyclability of medical waste. Developing countries must also prioritize improving waste management processes to maximize collection and recycling efforts while preventing plastic mismanagement, littering, and the associated risks of contamination and plastic leakage. In the European Union, both the "Waste Framework Directive" and the "Packaging and Packaging Waste Directive" have been revised by the European Parliament, establishing new objectives for bio-based feedstocks and recycling to support sustainable plastics production and more effective waste management (EC 2018). These strategies should be implemented and encompass the projected expansion of disposal facilities, the improvement of infrastructure for recycling, composting, landfilling, and incineration, increased coordination between involved parties, and the empowerment of producers, retailers, and municipalities. When recycling is not feasible, waste should be utilized as feedstock or incinerated as waste-to-energy (Prata et al. 2019).

It is essential to explore alternative end-of-life solutions for plastic waste that cannot be effectively recycled, such as composites and highly degraded materials (Braungart et al. 2007). As discussed, the use of recycled plastics is hindered by the lower costs of virgin materials. To overcome this challenge, the implementation of taxes and the enforcement of minimum requirements for the utilization of recycled materials could prove effective (Singh and Ruj 2015). Additionally, proper disposal and sterilization techniques for end-of-life PPE should be considered. International cooperation must be encouraged, facilitating the sharing of knowledge and technology. The cultivation of scientific advancements is crucial, as the future of humanity relies on technological and scientific knowledge to overcome challenges. Therefore, coordination among stakeholders, academia, citizens, and the government is of paramount importance in addressing these formidable issues.

6 Concluding Remarks

The COVID-19 pandemic put immense strain on regular waste management practices, resulting in mismanaged strategies such as mobile incineration, direct landfills, and local burnings. As a result, even the improper disposal of just 1% of face masks amounted to over 10 million items, weighing approximately 40 metric tons. The pandemic also exacerbated issues related to littering and waterlogged plastic along the coast, posing an increased risk of microplastic pollution. The significant environmental footprints of plastic pollution during this crisis can be attributed to the high demand for SUPs, the increased use of PPE, the surge in medical waste generation, and the prioritization of incineration and landfilling. Furthermore, the commencement of mass vaccination campaigns further escalated the generation of plastic-based medical and household waste. However, prioritizing human health over environmental health has resulted in the postponement and even reversal of policies aimed at reducing SUPs (WWF 2020). Recycling effects slowed down during the pandemic, and the demand for recycled plastic materials decreased due to declining profit margins. Consequently, there was an increase in the environmental footprint of plastics. It is crucial to adopt circular economy approaches, strengthen recycling practices, and strictly enforce policies against plastic pollution. Companies must also take responsibility by reducing their reliance on virgin plastic and increasing the proportion of recycled plastic in their products to fulfill their corporate social and environmental obligations. Meeting the sustainable development goals set by the United Nations requires a concerted effort to protect the environment for a post-pandemic future. In all, the recommended approaches to plastic waste management must be implemented and encompass: (1) the projected expansion of disposal facilities, (2) the enhancement of infrastructures for recycling, composting, landfilling, and incineration, (3) increased coordination among all stakeholders involved, and (4) the empowerment of producers, retailers, and municipalities. In cases where recycling is not feasible, waste should be utilized as feedstock or incinerated as waste-to-energy.

Appendix 1

The COVID-19 pandemic presented significant challenges, while also offering an opportunity to devise effective solutions to address plastic pollution. To ensure the efficient management of infectious plastic-based waste from the pandemic, it is crucial to tackle various process-related challenges, including source identification and segregation. In recognition of this need, numerous international organizations, such as UNICEF, WHO, European Commission, Asian Development Bank, UN-Habitat, and United Nations Environment Programme, along with national health organizations and leading pollution monitoring bodies like the Ministry of Ecology and Environment (China), Occupational Safety and Health Administration (USA), Central Pollution Control Board (India), Ministry of Health and Indigenous Medical Services (Sri Lanka), National System for Environmental Protection (Italy), and Ministry of Health and Population (Nepal), have issued comprehensive guidelines and advisories for the proper handling of COVID-19-related biomedical and household wastes as summarized in Table 2.

Table 2 Different guidelines issued by various organizations and agencies for the management of plastic-based COVID-19 waste (adapted from Das et al. 2020)

Organization	Policy and recommendation
International organizations	
Asian Development Bank (ADB 2020)	• COVID-19 Infectious solid waste (PPE included) should be wrapped twice before disposal/treatment • Recycling activity should be avoided for the treatment of household/medical infectious COVID-19 waste. It should be treated via sanitary landfilling or incineration
World Health Organization and UNICEF (WHO 2020)	• Disposable gloves, masks, and gowns should be disposed of as environmentally harmful waste after each use and not reused • Contagious Waste generated during patient care, including confirmed infection with COVID-19 (for example, PPE, pathological waste) must be collected safely in clearly marked lined containers and sharp, safe boxes, and then treated, preferably on-site and then safely disposed of • Waste generated in the house during the quarantine should be packed in strong black bags and closed completely before the disposal and final collection by municipal waste services • The most preferred treatment options for this infectious waste are autoclaving or a high-temperature dual-chamber incinerator

(continued)

Table 2 (continued)

Organization	Policy and recommendation
Union Nations Environment Program (UNEP 2020)	• Efficient disposal of healthcare and biomedical waste requires proper identification, separate collection, segregation, storage, transport, handling, and disposal under the hygiene and protection of personnel • Contaminated medical waste should not be mixed with domestic waste; it must be collected by municipalities and waste disposal companies and disposed of separately from the other household stream and treated as household waste
UN-Habitat (UN-Habitat 2020)	• All potentially infectious waste should be placed in a colored bag (double bag if possible) tied with string or tape. They should be handled as residual waste not intended for material recovery • Temporary local storage and heat treatment of potentially infectious waste of all the identified waste needs to be enforced • When infectious waste cannot be heat treated, an appropriate and safe landfill should be put in place • Enhance the frequency of waste collection and redesign the assets for waste collection based on the mapping of waste generation sources • Before the heat treatment of infectious waste, it should be stored temporarily on-site for 72 h • Recycling of materials in multi-material recovery systems can be performed after proper storage of materials and subsequent sterilization
National organizations	
Ministry of Ecology and Environment (MOEE 2021)	• Management required for emergency disposal of hazardous COVID-19 waste should be a priority for medical waste disposal facilities • Portable medical disposal treatment facilities such as hazardous waste incineration, industrial furnaces, and other facilities should be carried out in consultation with the relevant medical authorities • Personal hygiene should be practiced while handling (storage, transport, and disposal) COVID-19 waste

(continued)

Table 2 (continued)

Organization	Policy and recommendation
Central Pollution Control Board (CPCB 2020)	• COVID-19 waste should be kept in two-layer plastic bags, and a separate color-coded bin must be used in isolation and COVID-19 intensive care center for proper segregation of waste • Bags/containers used for collecting COVID-19 waste should be labeled as "COVID-19 waste." • The waste handler should maintain good hygiene when dealing with infectious COVID-19 waste, such as used PPE, masks, and gloves • Plastic-based waste should be sterilized, shredded, and then recycled • Non-recyclable plastic waste should go for incineration
Occupational Safety and Health Administration (OSHA 2020)	• Special attention should not be given to the COVID-19 contaminated waste. Waste management of contaminated waste should be the same as any other non-contaminated solid waste • Safe work practices such as the use of PPE, puncture-resistance gloves, eye and face protection must be adopted by workers to prevent exposure to infectious materials
National System for Environmental Protection (SNPA 2020)	• Based on the source of waste collection (COVID-19-impacted and non-impacted sources), contaminated plastic waste should be categorized • Hazardous COVID-19 waste should be collected in two-layer bags and treated separately
Ministry of Health and Indigenous Medical Services (MOHIMS 2020)	• It is recommended to segregate hazardous waste generated from the COVID-19 non-impacted and impacted sources and keep it in a color-coded container • During the transportation of infectious waste to the treatment plant, it is important to check for leakage or spillage to avoid contamination • Infectious waste, including plastic-based protective equipment, must be autoclaved or incinerated according to an appropriate safety protocol

References

ADB (2020) Managing infectious waste during the COVID-19 pandemic. Asian Dev Bank (ADB). https://www.adb.org/publications/managing-medical-waste-covid1

Adyel TM (2020) Accumulation of plastic waste during COVID-19. Science 369(6509):1314–1315

Allison AL, Ambrose-Dempster E, T Aparsi D, Bawn M, Casas Arredondo M, Chau C et al (2020) The environmental dangers of employing single-use face masks as part of a COVID-19 exit strategy

Bawaneh K, Ghazi Nezami F, Rasheduzzaman M, Deken B (2019) Energy consumption analysis and characterization of healthcare facilities in the United States. Energies 12(19):3775

Bengali S (2020) The COVID-19 pandemic is unleashing a tidal wave of plastic waste. The Los Angeles Times

Bermingham F, Tan S (2020) Coronavirus: China's mask-making juggernaut cranks into gear, sparking fears of over-reliance on world's workshop. South China Morning Post 12

Bown CB (2020) COVID-19: China's exports of medical supplies provide a ray of hope 2020, 26 March 2020. https://www.piie.com/blogs/trade-and-investment-policy-watch/covid-19-chinas-exports-medical-supplies-provide-ray-hope

Braungart M, McDonough W, Bollinger A (2007) Cradle-to-cradle design: creating healthy emissions – a strategy for eco-effective product and system design. J Clean Prod 15(13):1337–1348

Carus M (2018) Global markets and trends of bio-based building blocks and polymers 2018 – 2023. Nova Institute, Hurth, Germany. https://europeanbiotechnology.com/fileadmin/Content/NewsAndStories/2019/Nova_exec.pdf

Chanprateep S (2010) Current trends in biodegradable polyhydroxyalkanoates. J Biosci Bioeng 110(6):621–632

Chin AWH, Chu JTS, Perera MRA, Hui KPY, Yen H-L, Chan MCW et al (2020) Stability of SARS-CoV-2 in different environmental conditions. The Lancet Microbe 1(1):e10

Christopher Weikart RL (2020) A new hybrid material for packaging covid-19 vaccines and other biologics: combining the best of glass and plastic. Am Pharmaceut Rev 19–24

Cirella GT (2022) Human settlements: urbanization, smart sector development, and future outlook. Springer, Singapore

Cirella GT (2020) Sustainable human-nature relations: environmental scholarship, economic evaluation, urban strategies. Springer, Singapore

Cirella GT, Zerbe S (2014) Quizzical societies: a closer look at sustainability and principles of unlocking its measurability. Int J Sci Soc 5:29–45. https://doi.org/10.18848/1836-6236/CGP/v05i03/51427

Cirella GT, Tao L, Mohamed S (2007) An application of an adaptive quantitative method to measure the sustainability of the gold coast, Australia. J Coast Res SI50:52–56

Cirella GT, Wanjiku S, Paczoski A, Tiruneh S (2020) Human-nature relations: the unwanted filibuster. In: Cirella GT (ed) Sustainable human-nature relations: environmental scholarship, economic evaluation, urban strategies. Springer, Singapore, pp 3–22

Cirella GT, Russo A, Benassi F, Czermanski E, Goncharuk A, Oniszczuk-Jastrząbek A (2021) Energy re-shift for an urbanizing world. Energies 14:5516. https://doi.org/10.3390/en14175516

Corburn J, Vlahov D, Mberu B, Riley L, Caiaffa WT, Rashid SF et al (2020) Slum health: arresting COVID-19 and improving well-being in urban informal settlements. J Urban Health 97(3):348–357

CPCB (2020) Guidelines for handling, treatment and disposal of waste generated during treatment/diagnosis/quarantine of COVID-19 patients: CPCB. http://cpcbenvis.nic.in/pdf/1595918059_mediaphoto2009.pdf

Cutler S (2020) Mounting medical waste from COVID-19 emphasizes the need for a sustainable waste management strategy

Das KP, Sharma D, Saha S, Satapathy BK (2021) From outbreak of COVID-19 to launching of vaccination drive: invigorating single-use plastics, mitigation strategies, and way forward. Environ Sci Pollut Res Int 28(40):55811–55845

Data OWi (2022) Share of people vaccinated against COVID-figure: our world in data. https://ourworldindata.org/covid-vaccinations
EC (2020) Waste management in the context of the coronavirus crisis 2020. https://ec.europa.eu/info/sites/default/files/waste_management_guidance_dg-env.pdf
Ecochain (2020) The rise of the face mask: what's the environmental impact of 17 million N95 masks? Ecochain. ecochain.com/knowledge/footprint-face-masks-comparison
European Parliament (2018) Proposal for a directive of the European parliament and of the council on the reduction of the impact of certain plastic products on the environment, European Parliament. https://www.europarl.europa.eu/doceo/document/A-8-2018-0317_EN.html
Fadare OO, Okoffo ED (2020) Covid-19 face masks: a potential source of microplastic fibers in the environment. Sci Total Environ 737:140279
FDA (2020) Coronavirus testing basics 2020. www.fda.gov/consumers/consumer-updates/coronavirus-testing-basics
FluoroCarbon (2021) What is polychlorotrifluoroethylene? (PCTFE): FluoroCarbon. https://fluorocarbon.co.uk/news-and-events/post/11/what-is-polychlorotrifluoroethylene-pctfe
Fu Y, Yang X, Wang T, Supriyadi A, Cirella GT (2022) Spatial pattern characteristics of the financial service industry: evidence from Nanjing, China. Appl Spat Anal Policy 15. https://doi.org/10.1007/s12061-021-09409-z
Gallego-Schmid A, Mendoza JMF, Azapagic A (2019) Environmental impacts of takeaway food containers. J Clean Prod 211:417–427
Hatti-Kaul R, Nilsson LJ, Zhang B, Rehnberg N, Lundmark S (2020) Designing biobased recyclable polymers for plastics. Trends Biotechnol 38(1):50–67
Heidari M, Garnaik PP, Dutta A (2019) 11 - The valorization of plastic via thermal means: industrial scale combustion methods. In: Al-Salem SM (ed) Plastics to energy. William Andrew Publishing, pp 295–312
Hoornweg DB-T, Perinaz (2012) Urban development series; knowledge papers no. 15. World Bank, Washington, DC. What a waste : a global review of solid waste management Washington: World Bank [knowledge papers no. 15.]. https://openknowledge.worldbank.org/handle/10986/17388
Imbert E, Ladu L, Tani A, Morone P (2019) The transition towards a bio-based economy: a comparative study based on social network analysis. J Environ Manag 230:255–265
IMD (2020) There will be no COVID-19 vaccine without the right packaging: IMD. https://www.imd.org/news/updates/no-covid-vaccine-without-right-packaging/
Ioannidou SM, Pateraki C, Ladakis D, Papapostolou H, Tsakona M, Vlysidis A et al (2020) Sustainable production of bio-based chemicals and polymers via integrated biomass refining and bioprocessing in a circular bioeconomy context. Biores Technol 307:123093
Jiao G, Lu L, Chen G, Huang Z, Cirella GT, Yang X (2021) Spatiotemporal characteristics and influencing factors of tourism revenue in the Yangtze River delta urban agglomeration region during 2001–2019. Sustainability 13:3658. https://doi.org/10.3390/su13073658
Jribi S, Ben Ismail H, Doggui D, Debbabi H (2020) COVID-19 virus outbreak lockdown: what impacts on household food wastage? Environ Dev Sustain 22(5):3939–3955
Kahraman C, Orobello C, Cirella GT (2022) Changing dynamics with COVID-19: future outlook. In: Cirella GT (ed) Human settlements: urbanization, smart sector development, and future outlook. Springer, Singapore, pp 235–252
Karliner J, Slotterback S, Boyd R, Ashby B, Steele K (2019) Health care's climate footprint. Health Care Without Harm and ARUP
Karliner J, Slotterback S, Boyd R, Ashby B, Steele K, Wang J (2020) Health care's climate footprint: the health sector contribution and opportunities for action. Eur J Public Health 30(Supplement_5)
Kimini A (2020) How the COVID-19 plastic boom could save the oil industry. OilPrice com
Klemeš JJ, Van Fan Y, Jiang P (2020a) The energy and environmental footprints of COVID-19 fighting measures–PPE, disinfection, supply chains. Energy 211:118701

Klemeš JJ, Van Fan Y, Tan RR, Jiang P (2020b) Minimising the present and future plastic waste, energy and environmental footprints related to COVID-19. Renew Sustain Energy Rev 127:109883

Klemeš JJ, Fan YV, Jiang P (2021) Plastics: friends or foes? The circularity and plastic waste footprint. Energy Sour Part a: Recov Utiliz Environ Effects 43(13):1549–1565

Konda A, Prakash A, Moss GA, Schmoldt M, Grant GD, Guha S (2020) Aerosol filtration efficiency of common fabrics used in respiratory cloth masks. ACS Nano 14(5):6339–6347

Lewandowski R, Cirella GT (2022) Performance management systems: trade-off between implementation and strategy development. Oper Manag Res 16. https://doi.org/10.1007/s12063-022-00305-4

Lispasto (2017) Unit emissions 2017. lipasto.vtt.fi/yksikkopaastot/indexe.htm.

Markets RA (2020) COVID-19 impact on packaging market by material type, application and region—global forecast to 2021. Research and Markets, Dublin, Ireland. https://www.shorturl.at/knxBO

Martínez Silva P, Nanny MA (2020) Impact of microplastic fibers from the degradation of nonwoven synthetic textiles to the Magdalena river water column and river sediments by the city of Neiva, Huila (Colombia). Water 12(4):1210

McCarthy N (2020) The countries with the most critical care beds per capita. Forbes

MOEE (2021) Management and technical guidelines for emergency disposal of medical waste infected by the novel coronavirus pneumonia epidemic: Ministry of Ecology and Environment 2021. http://www.fao.org/faolex/results/details/fr/c/LEX-FAOC197697/

MOHIMS (2020) Guidelines for handling, treatment, and disposal of waste generated during treatment/diagnosis/quarantine of COVID-19 patients: Ministry of Health. https://cpcb.nic.in/uploads/Projects/Bio-Medical-Waste/BMWGUIDELINES-COVID_1.pdf

Mu-Hyun C (2020) ZDNet. https://www.zdnet.com/article/korea-sees-steep-rise-in-online-shopping-during-covid-19-pandemic/].

Ncube LK, Ude AU, Ogunmuyiwa EN, Zulkifli R, Beas IN (2020) Environmental impact of food packaging materials: a review of contemporary development from conventional plastics to polylactic acid based materials. Materials 13(21):4994

Oceans Asia (2020) No shortage of surgical mask at the beach 2020. http://oceansasia.org/beach-mask-coronavirus/

Oniszczuk-Jastrząbek A, Czermanski E, Cirella GT (2020) Sustainable supply chain of enterprises: value analysis. Sustainability 12:su12010419. https://doi.org/10.3390/su12010419

OSHA (2020) Control and prevention: OSHA. https://www.osha.gov/coronavirus/control-prevention

Phadke R, dos Santos Costa AC, Dapke K, Ghosh S, Ahmad S, Tsagkaris C et al (2021) Eco-friendly vaccination: tackling an unforeseen adverse effect. J Clim Change Health 1:100005

Potting J, Hekkert M, Worrell E, Hanemaaijer A (2017) Circular economy: measuring innovation in the product chain: PBL Publishers

Prata JC, Silva ALP, da Costa JP, Mouneyrac C, Walker TR, Duarte AC et al (2019) Solutions and integrated strategies for the control and mitigation of plastic and microplastic pollution. Int J Environ Res Public Health 16(13):2411

Prata JC, Silva ALP, Walker TR, Duarte AC, Rocha-Santos T (2020) COVID-19 pandemic repercussions on the use and management of plastics. Environ Sci Technol 54(13):7760–7765

Rattner N (2020) As coronavirus restrictions drag on, Americans shift online spending from stockpiling to entertainment: CNBC. https://www.cnbc.com/2020/04/19/coronavirus-what-americans-are-buying-online-while-in-quarantine.html

Saadat S, Rawtani D, Hussain CM (2020) Environmental perspective of COVID-19. Sci Total Environ 728:138870

Schnurr REJ, Alboiu V, Chaudhary M, Corbett RA, Quanz ME, Sankar K et al (2018) Reducing marine pollution from single-use plastics (SUPs): a review. Mar Pollut Bull 137:157–171

Silva ALP, Prata JC, Walker TR, Campos D, Duarte AC, Soares AM et al (2020) Rethinking and optimising plastic waste management under COVID-19 pandemic: Policy solutions based

on redesign and reduction of single-use plastics and personal protective equipment. Sci Total Environ 742:140565
Singh RK, Ruj B (2015) Plasticwaste management and disposal techniques - Indian scenario. Int J Plast Technol 19(2):211–226
SNPA (2020) COVID-19 emergency: SNPA indications on waste management, SNPA. https://www.snpambiente.it/2020/03/24/emergenza-covid-19-indicazioni-snpa-sulla-gestione-dei-rifiuti/
Spring C, Cirella GT (2022) Fostering sustainable development: green energy policy in the European Union and the United States. Human settlements: urbanization, smart sector development, and future outlook. Springer, Singapore, pp 101–137
Stokes G (2020) No shortage of surgical mask at the beach. Oceans Asia
Tobías A (2020) Evaluation of the lockdowns for the SARS-CoV-2 epidemic in Italy and Spain after one month follow up. Sci Total Environ 725:138539
Tripathi A, Tyagi VK, Vivekanand V, Bose P, Suthar S (2020) Challenges, opportunities and progress in solid waste management during COVID-19 pandemic. Case Stud Chem Environ Eng 2:100060
UNEP (2018) Single-use plastics: a roadmap for sustainability: UN Environment Program
UNEP (2020) Waste management an essential public service in the fight to beat COVID-19. UNEP, United Nation Environment Program. https://www.unep.org/news-and-stories/press-release/waste-management-essential-public-service-fight-beat-covid-19
UN-Habitat (2020) UN Habitat : response to COVID-19 : strategic guidance on solid waste management: UN-Habitat. https://unhabitat.org/un-habitat-covid-19-response-plan
Walensky RP, Del Rio C (2020) From mitigation to containment of the COVID-19 pandemic: putting the SARS-CoV-2 genie back in the bottle. JAMA 323(19):1889–1890
WHO (2020) Water, sanitation, hygiene, and waste management for SARS-CoV-2, the virus that causes COVID-19: WHO. https://www.who.int/publications/i/item/WHO-2019-nCoV-IPC-WASH-2020.4
WHO (2021) Sustainability in vaccine packaging WHO. https://www.who.int/immunization/policy/committees/Sustainability_in_Vaccine_Packaging_May11_DRAFT.pdf
Windfeld ES, Brooks MS-L (2015) Medical waste management – a review. J Environ Manag 163:98–108
WWF Italy (2020) Nello smaltimento di macherine e guanti serve resposabilità 2020. https://www.wwf.it/scuole/?53500%2FNello-smaltimento-di-mascherine-e-guanti-serveresponsabilita
WWF (2020) In the disposal of masks and gloves, responsibility is required WWF International, June 2020, 8
Xanthos D, Walker TR (2017) International policies to reduce plastic marine pollution from single-use plastics (plastic bags and microbeads): a review. Mar Pollut Bull 118(1):17–26
Xiao Y, Torok ME (2020) Taking the right measures to control COVID-19. Lancet Infect Dis 20(5):523–524
Zehrung D, Jarrahian C, Giersing B, Kristensen D (2017) Exploring new packaging and delivery options for the immunization supply chain. Vaccine 35(17):2265–2271
Zhu Y, Romain C, Williams CK (2016) Sustainable polymers from renewable resources. Nature 540(7633):354–362

Urbanization and CO_2 Emissions: Panel Data Analysis of EU Countries

Jana Chovancová, Igor Petruška, Giuseppe T. Cirella, and Eva Litavcová

Abstract Recognizing the recent pace of urbanization and carbon emissions growth, this study investigates the dependence of CO_2 emissions on the urbanization of European Union (EU) countries, economic growth, total energy consumption, the amount of energy produced from renewable sources, and nuclear energy as low-carbon energy source. The main questions addressed in this study are: (1) Are there long-term and short-term relationships between the variables examined? (2) What is the connection between carbon emissions and the independent variables at different periods? (3) Can urbanization, economic development, total energy consumption, and the share of renewable and low-carbon energy sources predict the evolution of carbon emissions and thus the achievement of the EU's carbon neutrality targets? We analyze a sample of 22 European countries for the period 1992–2019. Concerning the COVID-19 pandemic that hit Europe in early 2020, this study can be used as a comparative sample and can serve to assess the impacts of the pandemic on the evolution of greenhouse gas emissions during the pandemic as well as after the pandemic has subsided and economies have restarted. The models used in the study are as follows: threshold cross-sectional model, panel threshold models with threshold variable (urbanization of the country), Panel Models (Fixed Effects (FE), Random Effects, Least Squares Dummy Variables), Fully Modified OLS, Dynamic OLS, Panel VAR model, Pooled Mean Group Estimation, Mean Group Estimation, and Dynamic FE estimation. General moments methods (GMM) were used in the estimation of short-run relationship dynamics. The impulse response functions were

J. Chovancová (✉) · I. Petruška
Faculty of Management and Business, University of Presov, 080 01 Prešov, Slovakia
e-mail: jana.chovancova@unipo.sk

I. Petruška
e-mail: igor.petruska@unipo.sk

G. T. Cirella
Faculty of Economics, University of Gdansk, 81-824 Sopot, Poland
e-mail: gt.cirella@ug.edu.pl

E. Litavcová
Department of Mathematics, Faculty of Education, Catholic University in Ružomberok, 034 01 Ružomberok, Slovakia
e-mail: eva.litavcova@ku.sk

G. T. Cirella and B. Dahiya (eds.), *City Responses to Disruptions in 2020*, Advances in 21st Century Human Settlements, https://doi.org/10.1007/978-981-99-7988-2_8

used to analyze the results obtained in the GMM model. The Stata 15.1 econometric software environment was used to test unit root, variable cointegration, and to estimate the models. The best-fitting models were selected from the above models using statistical tests. Understanding the relationship between these variables is essential for informed and evidence-based decision-making and the adoption of new or revision of existing policies and strategies promoting the carbon–neutral and green economy at the EU and national levels.

Keywords Urbanization in Europe · CO_2 emission levels · Economic growth · Energy consumption · Econometrics

1 Introduction

Urbanization in Europe is a persistent phenomenon in terms of urban sprawl and the growth rate of the population living in the cities and urban areas. In 2021, the urbanization rate in Europe reached 75% and this trend is expected to increase to approximately 83.7% in 2050 (The World Bank 2020; United Nations 2019). Cities interact with and influence the territory that surrounds them, often significantly affecting the environment of much larger areas. Their development is also driven by external factors such as demographic change, the need for mobility and energy supply, globalization, and climate change. Cities, large and small, are often described as engines of progress—often driving much of our cultural, intellectual, educational, and technological success and innovation. Thanks to a high concentration of labor, technology, and infrastructure, cities make a significant contribution to economic growth, with up to 80% of gross domestic product (GDP) generated in cities (Dobbs et al. 2011). However, the current trend toward new approaches to urban development in the EU, characterized by low population density, is leading to higher energy and resource consumption and greater demands on transport and land. Meeting the growing energy needs associated with rising living standards is also linked to an increase in greenhouse gas emissions, as well as other environmental and social problems.

Urbanization is related to several factors, as an urban population has significantly different consumption patterns from a rural population. Urban populations are more exposed to adverse phenomena associated with the urban environment, such as air pollution and associated health problems, and reduced quality of life (Huttmanová et al. 2019). Rapid urbanization puts pressure on energy systems, water supplies, sanitation, infrastructure, transport, health systems, and the environment. Concerning some of the negative aspects that have accompanied urbanization, the United Nations Sustainable Development Goals have been supplemented by Goal No. 11 "Sustainable cities and communities", aiming to make cities and human settlements inclusive, safe, resilient, and sustainable (United Nations 2015).

Cities also play an important role in mitigating and adapting to climate change. For several decades, urban policies have mainly focused on issues of poverty, crime, inadequate infrastructure, and urban decay. Recently, environmental issues and climate change are gaining prominence. Greater emphasis on these issues stems from the recognition that cities are heavily involved in and affected by climate change. According to UN estimates, cities consume about 75% of global primary energy, making them a key contributor to climate change, responsible for 60% of global carbon dioxide (CO_2) emissions, with transport and buildings being among the major sources of greenhouse gasses (UN Habitat 2020). Several studies and climatic experts from IPCC in their latest report confirm that the world faces unavoidable multiple climate hazards over the next two decades with the average global temperature exceeding 1.5 °C (i.e., compared to pre-industrial level). If no mitigation action is taken, warming will further aggravate the impacts, some of which will be irreversible (IPCC 2022; Lu et al. 2018; Malhi et al. 2020). On the other hand, densely populated cities are highly vulnerable to the adverse effects of climate change. Rising global temperatures cause sea levels to rise, increasing the number of extreme weather events such as floods, storms, and hot extremes including heat waves associated with urban heat islands. Multiple climate change-related hazards have a negative impact on the health of its inhabitants, with those economically and socially marginalized as the most vulnerable (Qian et al. 2022; Oniszczuk-Jastrząbek et al. 2020; Spring and Cirella 2022; Voelkel et al. 2018). Extreme and slow-onset events can damage or destroy key infrastructure including transportation, water, sanitation, and energy systems, resulting in economic losses, disruptions of services, and impacts to well-being.

It is increasingly clear that climate change and urbanization are intricately interrelated questions at the level of both the effect of climate change and the cause of that change itself. Therefore, it is very important to understand the relationship between these megatrends as well as their relationship to other variables such as economic growth and energy consumption that are directly related to them. The present study, therefore, examines the relationship between the rate of urbanization and CO_2 emissions as a major greenhouse gas contributing to climate change. At the same time, it examines the impact of energy consumption and economic growth on the development of CO_2 emissions, as well as the mitigation effect associated with renewable and low-carbon energy sources. To assess the relationship between these variables, the panel VAR model, Pooled Mean Group Estimation (PMG), Mean Group Estimation (MG), Dynamic FE estimation, and the Granger causality test were applied. The short-run relationship dynamics were estimated by General moments methods (GMM). The results of the analysis provide an important knowledge base for informed and responsible decision-making and for creating and modifying urban and development policies toward a more sustainable and resilient society.

2 Literature Review

The complexity and dynamics of urbanization, economic growth, energy consumption, and CO_2 emissions as well as the relationships between them are reflected in the vast number of studies carried out in this area. Studies in the field of energy economics deal with the relationship between economic growth and energy consumption, not taking into account environmental impacts. On the other hand, studies in the field of environmental economics focus on the relationship between economic growth, and/or energy consumption, and the environmental damage or growth of pollution (e.g., CO_2 emissions). To facilitate orientation in the given issue, the following literary review is divided into several sections dealing with the relationships between individual variables.

2.1 Economic Growth and the CO_2 Emissions Nexus

The relationship between economic growth and the state of the environment has been the subject of extensive discussions for a long time (Arrow et al. 1996; Ayres and van den Bergh 2005; Panayotou 2003; Stern 2004; Stern et al. 1996). Several authors argue that continuous economic growth in the limited conditions of the biosphere is not possible, and therefore the use of material and energy resources to achieve economic growth cannot last forever (e.g., Daly 2007). Others, e.g., Grossman and Krueger (1995) argue that economic growth only worsens the environment when the economy is based on heavy, polluting industries and that as the economy grows, the structure of the economy shifts to less polluting production and services and environmental impacts cease to grow and at some point, it will even begin to decline. This idea was formulated as a hypothesis of the Environmental Kuznets Curve (EKC), the conceptual anchoring of which was initiated by research on the relationship between economic growth and income inequality (Kuznets 1955). Kuznets assumed that as the economy develops, market forces first increase and then reduce overall economic inequality. According to the EKC hypothesis, economic growth first leads to deterioration (degradation) of the environment, but at a certain point economic growth is reached (the turning point), environmental degradation is reduced and the relationship between economic growth and the environment is improving. The EKC is based on the idea that environmental damage is for some time considered an acceptable side effect of economic growth (Arrow et al. 1995), but after reaching a certain level of per capita income, the pressure on the quality of the environment increases (Grossman and Krueger 1995). The curve has an inverted U shape with an economic indicator (e.g., GDP) on the horizontal axis and an environmental indicator on the vertical axis.

The use of EKC can be observed on two levels: theoretical and empirical. Theoretical studies devoted to the EKC began in the early 1990s. One of the first was a study on measures to improve air quality (Grossman and Krueger 1991). Other theoretical works from this period include Panayotou (1994) and Shafik (1994), in which the authors provide the earliest and most detailed explanation of the possible relationship between the degree of environmental degradation and the level of economic development expressed by the EKC (Yandle et al. 2004). The empirical research of EKC focused mainly on two areas. The first group consists of studies examining the validity of the EKC for various environmental indicators (e.g., Bo (2011), De Bruyn et al. (1998), Hilton and Levinson (1998), Kaika and Zervas (2013), Selden and Song (1994), and Timmons Roberts and Grimes (1997)). These studies confirm the EKC hypothesis that although economic growth leads to an increase in emissions at the beginning of economic development, over time these emissions may decrease due to various technological and structural changes in the economy. The second group consists of studies determining the turning points for specific pollutants and countries (Apergis and Ozturk 2015; Gawande et al. 2000; He and Richard 2010; Özokcu and Özdemir 2017; Park and Lee 2011; Patel et al. 1995; P. Wang et al. 1998; Zaim and Taskin 2000). Numerous studies employing extensive materials and methods empirically demonstrate the validity of the EKC for various pollutants and processes in the environmental sphere (i.e., air, water, deforestation), as well as for economic sectors and countries. Recognizing that many conclusions confirm the validity of the EKC, several authors have applied this approach to complex international research (Bölük and Mert 2014; Cole et al. 1997; Miglietta et al. 2017; Sarkodie and Strezov 2018; Shafiei and Salim 2014; Waqih et al. 2019).

The validity of the EKC theory has been confirmed for some types of emissions (e.g., sulfur dioxide), but for many indicators, including greenhouse gas emissions, there is ample evidence of empirical studies that do not support this theory, e.g., Bölük and Mert (2014), Galeotti and Lanza (2005), and Huang et al. (2008). Recent works by respected authors examined the existence of EKC hypothesis in different economies (Al-Mulali and Che Sab 2018; Al-Mulali and Ozturk 2016; Nicholas Apergis et al. 2017; Bilgili et al. 2016; Dogan and Turkekul 2016; Kais and Ben Mbarek 2017; Katircioğlu and Katircioğlu 2018; Olale et al. 2018; Ozatac et al. 2017; Özokcu and Özdemir 2017; Zoundi 2017). In addition to the findings of these studies, some researchers find empirical evidence to support the N-shaped EKC hypothesis that the growth of income first worsens the environment and then reduces pollution, but increases environmental pressures once a certain level of economic growth is reached (Alam et al. 2011; Balcilar et al. 2014). In addition to the classic inverted U-shaped and N-shaped curve, the researchers also reported U-shape, inverted N-shape, M-shape, and a monotonically increasing or decreasing relationship between different greenhouse gas (GHG) emissions and economic growth. For example, Bekhet and Othman (2018) provide evidence supporting the EKC inverted N-type hypothesis for Malaysia between 1971 and 2015. Furthermore, a monotonically growing relationship between CO_2 and income can be found in a study by Akbostanci et al. (2009) for 58 provinces in Turkey. In addition to these findings, Yang et al. (2015) analyze the EKC hypothesis for 67 countries from 1971 to 2010 and find evidence supporting inverted N, M, inverted U, and monotonically growing shape, which vary depending on different regions or economic development.

2.2 Economic Growth and the Energy Consumption Nexus

The energy-growth nexus is emphasizing the cointegration between economic growth and energy consumption. Many country-specific studies were carried out to examine the relationships between the two variables to identify four types of energy-growth nexus: (1) the growth hypothesis is valid when there is a unidirectional causality running from energy consumption to economic growth, i.e., an increase in energy consumption contributes to economic growth, (2) the conservation hypothesis implies that there exists unidirectional relationship running from economic growth to energy consumption, (3) the feedback hypothesis considers a two-way causal link between energy consumption and economic growth, i.e., both variables are interdependent, and (4) the neutrality hypothesis states that there are no dynamic links between the two variables. For example, Apergis and Payne (2009) reported short-run and long-run causality running from energy consumption and economic growth in six Central American countries, which supports the growth hypothesis. (Ozturk et al. 2010) in their study revealed the existence of bi-directional short-run causality between natural gas consumption and income level supporting the feedback hypothesis in six Gulf Cooperation Countries. The study by Fang and Chang (2016) examines the cointegration of energy consumption and economic growth in 16 Asia–Pacific countries. Results provide evidence of a causal relationship running from economic growth to energy use, however, the relationship varies for individual countries.

Saidi and Hammami (2015) performed global analysis of 58 countries, finding that a statistically significant positive impact of economic growth on energy consumption was present only in the case of four global panes, including Europe, Asia, and the Latin American and Caribbean region. Long-run unidirectional causality running from economic growth to energy consumption was found in the case of Bangladesh, both at the aggregate and sectoral levels. The results suggest that the economic growth of the country is not energy-dependent, which is in line with the conservation hypothesis (Amin and Alam 2018). This is in agreement with the findings of Oh and Lee (2004) for Korea, where long-run unidirectional causal relation running from GDP to energy consumption was confirmed. However, the opposite findings are recorded in the study by Kwakwa (2012) for Canada, and Shiu and Lam (2004) for China, confirming the unidirectional causality running from energy consumption to GDP, but not vice versa.

The causality of the two variables was scrutinized in the case of the United States at a country as well as sectoral level in the study of Kourtzidis et al. (2018). The findings support the neutrality hypothesis for all sectors. The only exception is total energy consumption at the country level where exists unidirectional causality running from energy consumption to economic growth, supporting the growth hypothesis. Menegaki and Tugcu (2017) studied the causality between energy consumption and sustainable economic welfare growth, as an alternative of commonly used GDP, in G7 countries. The study provides evidence that G7 economies can reduce energy consumption without compromising sustainable economic welfare in the long run. The energy-growth nexus in OECD countries was analyzed in the study of Gozgor et al. (2018). The results provide evidence of the positive effect of renewable as well as non-renewable energy consumption on economic growth, confirming the validity of the growth hypothesis for 69 examined countries. Omri (2014) performed an international literature review on the nexus between economic growth and four types of energy consumption: total energy consumption, electricity consumption, nuclear energy consumption, and renewable energy consumption. The analysis brings mixed results confirming the quadruplet hypotheses mentioned above. More recently, Ahmad et al. (2020) performed a literature review of the 50 most prolific and frequently cited articles researching energy-growth nexus. The review focuses on a type of analysis and methodological apparatus used in reviewed studies as well as outcomes of the studies in terms of the type of relationship between energy consumption and economic growth.

2.3 *Economic Growth, Energy Consumption, and the CO_2 Emissions Nexus*

In light of the escalating effects of climate change, the interrelationship between economic growth, energy consumption, and greenhouse gas emissions is gaining importance. This is reflected in the number of studies that have been carried out in this area in different countries and regions. The findings are mixed across countries, periods, and research approaches. For example, Ang (2007) focused his scientific interest on the economy–energy–emissions nexus in France. Empirical results over the period 1960–2000 suggest a long-run quadratic relationship between CO_2 emissions and economic growth and more energy use results in more CO_2 emissions. Zhang (2016) explored the trio nexus in China, including environmental regulation as the next variable. The results confirm the existence of long-term equilibrium among examined variables. The findings indicate that energy consumption positively impacts economic growth, however, CO_2 emissions are positively related to the economic slowdown. Environmental regulation plays an important role as it boosts the economy; however, results are visible only in the long run.

A geographically wide analysis was performed by Chen et al. (2016) for 188 countries. Results confirm the existence of long-run relationships between economic growth, energy consumption, and CO_2 emissions for all countries. The authors point out the significant differences between developed and developing countries. While energy consumption negatively affects economic growth from a global perspective in the case of developing countries, the opposite is true for developed countries. The study also confirmed unidirectional causality from energy consumption to CO_2 emissions for both developed and developing countries.

Another multi-country analysis was performed by Arminen and Menegaki (2019). They found a bidirectional causal relationship between economic growth and energy consumption, confirming the feedback hypothesis. However, results do not support the EKC hypothesis, suggesting that 67 high- and upper-middle-income countries have not yet reached the turning point in the harmful effect of economic growth on the environment. The study by Mensah et al. (2019) analyzed 22 African countries, divided into oil exporters and non-oil exporters. A bidirectional link was found between energy consumption and economic growth and energy consumption and carbon emissions for all panels. However, there exist differences among studied groups: in the case of non-oil exporters, there exists unidirectional causality running from carbon emissions to economic growth in the long and short run, but in the case of oil-exporting countries, there exists a bidirectional link between variables in a long run.

More recently, Zhang et al. (2019) focused their scientific interest in the economy-carbon-energy nexus on 50 typical developing countries. The results of their study reveal the long-run equilibrium between carbon emissions and economic-energy variables. However, the regression results illustrate that the EKC theory of the inverted U shape curve is valid only for 7 countries. In the majority of developing countries, energy consumption has a positive and statistically significant impact on carbon emissions. The relationship between economic growth, energy consumption, and CO_2 emissions in 14 countries of the Danube region was explored in a study by Litavcová and Chovancová (2021). The results indicate that there exists a long-run cointegration between carbon emissions, energy consumption, and economic growth in four countries, i.e., Austria, Czechia, Slovakia, and Slovenia. However, in the short run, this impact is significant only in the case of Austria.

2.4 *Economic Growth and the Urbanization Nexus*

The literature provides ample evidence of the close relationship between economic growth, urbanization, and the sustainability concept. It is generally accepted that economic growth, often driven by industrialization, fosters the expansion of urban populations. Concurrently, urbanization plays an important role in economic growth, but it is crucial to consider the sustainability aspect (Cirella and Zerbe 2014a, b; Fu et al. 2022; Jiao et al. 2021; Lewandowski and Cirella 2022). The speed and nature of urbanization become critical factors in this equation. Sustainable urbanization practices can mitigate the negative environmental impacts associated with rapid urban growth, ultimately promoting a more balanced and environmentally responsible path of economic development. Thus, the interplay between economic growth, urbanization, and sustainability is central to shaping the future of our cities and their impact on the world (Russo and Cirella 2017; Cirella 2020, 2022; Cirella et al. 2020, 2021). Single-country analyses were performed for China (Bai et al. 2012; Liang and Yang 2019; Zheng and Walsh 2019). The studies confirm that urbanization has a causal effect on economic growth. Results indicate that urbanization boosts economic development via the accumulation of physical, knowledge, and human capital. The opposite direction of influence between the variables was found in the case of India (Ghosh and Kanjilal 2014), where the authors confirmed a unidirectional relationship going from economic growth to urbanization. However, the findings of these studies unanimously highlight the need for careful planning and management of the expansion of urban areas and massive migration to the cities, as this may jeopardize agriculture, land use, and food security and can be accompanied by adverse social impacts.

Studies examining cross-country panel contexts have a much wider presence in the literature. New emerging-market countries were examined by Bakirtas and Akpolat (2018). The bivariate analysis confirmed the panel Granger causality from urbanization to economic growth and energy consumption. In the panel of 7 ASEAN countries, Nguyen and Nguyen (2018) estimates the impact of urbanization on economic growth. Findings confirm the existence of a causal relationship between the two variables. Although urbanization has a positive effect on economic growth, this relationship is not linear. Once a threshold is reached, growth in urbanization may hamper economic growth. A negative relationship between urban population rates and economic growth was found in the study of Abdel-Rahman et al. (2006), which focused their interest on developing countries. Their observations show that it was the economic downturn in the country that led to a more massive migration of people living in rural areas to cities.

A panel of 170 countries was utilized in the study of Wang et al. (2018). Bidirectional Granger causality was present between economic growth and urbanization, as well as CO_2 emissions and energy consumption, CO_2 emissions and economic growth, CO_2 emissions and urbanization, energy consumption, and economic growth, both long run and short run. The results suggest that the interrelationships between the variables differ due to the economic maturity and income level of each sub-panel. However, examining the relationship between economic growth and urbanization is often set in a broader context. The authors are aware of the broader context that can enter and influence this relationship. Concerning sustainable development, therefore, many authors also include environmental variables in the model, such as CO_2 emissions (Armeanu et al. 2021; Liang and Yang 2019; Xu et al. 2019) or social factors, e.g., poverty (Nguyen et al. 2020). Several studies point out that rapid urbanization can have a negative impact on the environment in the form of extensive use of natural resources and energy (Han et al. 2018), air pollution (Shahbaz et al. 2016), ecosystem damage, biodiversity loss (Ali et al. 2019) but also in general to the quality of life of the population and their health (Cirella et al. 2007; Cirella and Tao 2008, 2009a, b; Cirella and Zerbe 2014a, b; Hanif 2018; Russo and Cirella 2020).

2.5 *Renewables and Nuclear Energy*

Mitigation of the adverse effects of climate change and the need to decarbonize economies is a frequently discussed topic, especially in the EU, which has set ambitious targets and aims to achieve carbon neutrality by the middle of the century. The energy dependence of many EU Member States on fossil fuels is also a strong driving force for deep decarbonization and energy transition. Efforts to decarbonize economies as well as to reduce the dependence on energy raw materials originating in third countries are associated with the possibility of higher use of renewable resources, such as wind, water, solar, geothermal energy, or so-called low-carbon sources, including nuclear energy. Here, however, the views of the professional community differ significantly. Some authors consider this resource as sustainable and inevitable for achieving decarbonization targets (Brook et al. 2014; Buongiorno et al. 2019; Czermański et al. 2020; Czermanski et al. 2021), while others point to the limitations of the technology (Pearce 2012) and the need to perceive technology throughout the life cycle lenses (Poinssot et al. 2014), pointing to the carbon footprint associated with power plant construction and the problem of nuclear waste (Yano et al. 2018).

However, the importance of renewables and nuclear energy in the study of the energy-economy-environment nexus is underlined by numerous studies that consider these factors as variables in model construction. For example, Cheng and Liu (2019) analyzed the effect of different energy sources' consumption on economic growth in China. Their findings indicate that in terms of multiplying effect, clean energy (i.e., hydro, nuclear, wind, and solar) has the highest contribution rate to economic growth. EKC hypothesis was confirmed for the United States, where an inverted U-shaped function was found between CO_2 and economic growth (Salari et al. 2021). The study indicates that renewable energy has a negative effect on CO_2 emissions, while total, non-renewable, industrial, and residential energy consumption has a positive impact on CO_2 emissions. Cai et al. (2018) investigated the relationship between clean energy consumption, economic growth, and CO_2 emissions in G7. They found unidirectional causality running from clean energy consumption to economic growth for the United States, Canada, and Germany and from carbon emissions to clean energy consumption for Germany.

Anwar et al. (2019) employed data for 59 countries divided into different income groups. The results vary among the studied panel countries. However, they conclude that a higher share of nuclear energy leads to a reduction of CO_2 emissions, except the panel of upper-middle-income countries. Radmehr et al. (2021) have analyzed the nexus between renewable energy consumption, CO_2 emissions, and economic growth in the EU countries in the period 1995–2014. They found feedback relation between economic growth and carbon emissions and between economic growth and renewable energy consumption. The findings also provide evidence on the unidirectional relationship between renewable energy and CO_2 emissions.

Single-country analysis was performed by Adebayo et al. (2022). They confirmed the negative effect of trade openness, economic growth, and renewable energy consumption on CO_2 emissions in Sweden. The evidence of the mitigation effect of renewable energy on carbon emissions was provided by Vural (2020) for the group of Sub-Saharan African countries. The findings reveal the positive effect of non-renewable energy and trade on carbon emissions, while the deployment of renewables is associated with CO_2 emissions reduction. Nuclear energy-economic growth in 13 OECD countries was examined by Ozcan and Ari (2017). Their findings support the feedback hypothesis, i.e., nuclear energy consumption and economic growth influence and complement each other, both in the short run and long run. Moreover, in 6 out of 13 countries, nuclear energy positively and significantly impacts the real

GDP in the long run. This analysis was complemented by Gozgor et al. (2018) who provide evidence that both non-renewable and renewable energy consumption in OECD countries is positively associated with a higher rate of economic growth.

Although several studies have addressed the interrelationship between economic growth, energy consumption, and CO_2 emissions, some of which include EU countries, there is a lack of analysis of recent data. Given the changing conditions of an exogenous nature, such as the COVID-19 pandemic and also the war in Ukraine, an analysis of the evolution of these factors is essential when deciding on a modification or complete overhaul of energy and climate policies as well as urban development policies. In addition, other important variables such as urbanization and renewable and low-carbon energy sources were included in the model design to determine their relevance for climate change mitigation.

3 Data and Methods

This study analyses the effect of energy consumption, renewable energy consumption, nuclear energy consumption, urban population, and GDP on CO_2 emissions. Data for 22 European countries from 1992 to 2019 were used for the analysis. The countries are as follows: Austria (AUT), Belgium (BEL), Bulgaria (BGR), Czech Republic (CZE), Denmark (DNK), Finland (FIN), France (FRA), Germany (DEU), Greece (GRC), Hungary (HUN), Ireland (IRL), Italy (ITA), Luxembourg (LUX), Netherlands (NLD), Poland (POL), Slovenia (SVN), Slovak Republic (SVK), Spain (ESP), Romania (ROU), Portugal (PRT), Sweden (SWE), and United Kingdom (GBR). The relationships between the four variables were examined. The list of investigated variables along with their abbreviated names used in the analyses is as follows:

- CO_2 emissions per capita (tons)—CO_2,
- urban population (% of total population)—Urban,
- GDP per capita (thousands USD)—GDP,
- total energy consumption per capita (MWh)—TEC,
- energy produced from renewable sources per capita (MWh)—RES, and
- energy produced in nuclear power plants per capita (MWh)—Nuclear.

3.1 Descriptive Statistics

The main object of interest is the effect of the urban population variable on CO_2 emissions per capita variable, as well as the effect of the other variables mentioned. In addition, the dependence of CO_2 emissions per capita on other variables was investigated in a multidimensional context. The variables (with the exception of nuclear power) were logarithmized before testing and estimating the models. Detailed descriptive statistics of each variable by country allow the reader to compare them.

The descriptive statistics of the variables in the whole set of n = 616 observations are presented in Table 1. Differences between variables across individual countries can be observed in the tables of descriptive statistics by country (Tables 2, 3 and 4), where countries are sorted in ascending order by the average of each of the variables. However, the statistical tables do not provide information on how the measured values are distributed in the min–max interval. Line charts were chosen to visualize the distribution of 28 values over time within each country (Figs. 1, 2, 3, 4, 5 and 6). An overview of the empirical probability (number) of observations in a set of all 616 observations for individual variables, using Kernel density estimation (KDE), is shown in Fig. 7.

Table 2 shows that Luxembourg has extremely high CO_2 emissions compared to the other 21 countries. Using the original data, it was possible to establish that Luxembourg's CO_2 levels have gradually decreased since 2005 until they reached their minimum in 2017 and have only slightly increased since then. Due to the large difference compared to other countries, Luxembourg is not included in Fig. 1. Figure 1 shows that all countries have tended to reduce their CO_2 emission levels since the second half of the first decade of the 3rd millennium and have at least maintained a stable level since the Russian-Ukrainian conflict in 2014. Nine out of 22 countries failed to go below 7 tons of CO_2 emissions per capita in 2019, the last year studied. After 2010, Finland, Ireland, Denmark, and Greece recorded a strong decrease in CO_2 emissions, while Austria and Bulgaria stagnated and Hungary even recorded an increase.

The evolution of the Urban variable, i.e., the proportion of urban population, is shown in Fig. 2. Figure 2 shows that the share of urban population in the Netherlands and Portugal has increased extremely strongly compared to the other countries throughout the study period. In several countries, the trend of the urban variable has changed since 2010 after the crisis, especially in Austria, where the trend went from a decrease to an increase. In Finland, Germany, Poland, Romania, and Slovakia, on the other hand, stagnation or decline was observed over the same period. Countries with a low share of urban population are Austria, Ireland, Romania, Slovenia, and Slovakia. These are the countries that did not exceed 70% during the entire reporting period. In contrast, the almost purely urban countries are Belgium, the Netherlands, and Luxembourg with an urbanization rate above 90% in the last years of the surveyed period. Sweden and Denmark are also approaching this value quite quickly.

Table 1 Descriptive statistics of investigated variables, all 616 observations

Variable	Mean	Std. dev	Min	Max
CO_2	8.784	3.721	3.818	31.253
Urban	72.093	12.462	49.130	98.041
GDP	30.406	17.058	4.504	124.591
TEC	44.345	17.233	17.920	114.632
RES	4.259	5.394	0.055	26.539
Nuclear	4.637	5.726	0	23.385

Table 2 Descriptive statistics of variables CO_2 and Urban, 28 observations for every country, ordered according to mean

CO_2					Urban				
Country code	Mean	Min	Max	Std. dev	Country code	Mean	Min	Max	Std. dev
ROU	4.605	3.818	5.705	0.607	SVN	52.027	50.543	54.822	1.422
PRT	5.491	4.599	6.697	0.671	ROU	53.597	52.780	54.227	0.427
HUN	5.553	4.444	6.143	0.555	SVK	55.313	53.726	56.719	1.079
SWE	5.692	4.188	7.145	0.931	PRT	57.698	49.130	65.764	5.077
FRA	6.385	4.971	7.280	0.779	AUT	59.222	57.115	62.417	1.672
BGR	6.493	5.665	7.357	0.440	IRL	60.433	57.293	63.405	1.885
ESP	6.705	5.386	8.394	0.966	POL	61.096	60.037	61.787	0.590
ITA	7.388	5.567	8.622	1.035	HUN	67.408	64.575	71.644	2.470
SVK	7.502	6.105	9.200	0.818	ITA	68.105	66.742	70.736	1.185
SVN	7.720	6.545	9.006	0.683	BGR	70.907	67.056	75.347	2.596
AUT	8.289	7.439	9.595	0.637	CZE	73.849	73.185	75.030	0.538
GBR	8.503	5.477	10.324	1.568	GRC	74.954	71.757	79.388	2.527
GRC	8.571	6.415	10.305	1.141	DEU	75.824	73.360	77.376	1.359
POL	8.719	7.961	9.800	0.499	FRA	77.332	74.401	80.709	1.971
DNK	9.480	5.557	14.240	2.342	ESP	77.633	75.610	80.565	1.611
IRL	9.913	7.602	12.391	1.482	GBR	80.348	78.172	83.652	1.846
NLD	10.539	9.056	11.704	0.699	NLD	82.189	70.223	91.876	7.127
DEU	10.607	8.405	12.079	0.901	FIN	83.097	80.128	85.446	1.683
FIN	10.859	7.529	13.900	1.731	SWE	84.918	83.361	87.708	1.271
BEL	10.954	8.628	12.721	1.544	DNK	86.148	84.898	87.994	1.108
CZE	11.604	9.450	13.969	1.267	LUX	86.561	81.528	91.223	3.154
LUX	21.680	15.628	31.253	4.388	BEL	97.383	96.542	98.041	0.449

As in the case of the CO_2 variable, the GDP variable in Luxembourg is also extremely high compared to other countries (Table 3). Using Luxembourg data, a sharp increase in values could be observed between a minimum of 33.755 € per capita in 1992 and a maximum of 124.591 € per capita in 2019, with only a very small decrease during the global crisis in 2009. The behavior of GDP during the period is similar in all other countries. Figure 3 shows that the 2008–2009 crisis affected more or less all countries. Greater problems with economic growth were later experienced by Greece after 2008, Italy between 2011 and 2015, and Slovakia from 2016 onwards. Ireland, on the other hand, has experienced a very strong increase in GDP since 2014.

In nine of the 22 countries surveyed, TEC grew by 2004 (Fig. 4). In 2005–2013, most countries tried to reduce TEC. After 2014, TEC increased more in most countries, followed by a subsequent slight decrease, but in Hungary, Poland, Romania,

Table 3 Descriptive statistics of variables GDP and TEC, 28 observations for every country, ordered according to mean

GDP					TEC				
Country code	Mean	Min	Max	Std. dev	Country code	Mean	Min	Max	Std. dev
BGR	12.322	5.748	25.312	5.880	ROU	20.514	17.920	24.051	1.766
ROU	13.470	4.504	33.340	8.566	PRT	26.595	20.794	28.849	2.315
POL	17.069	6.178	35.165	8.662	HUN	27.670	24.398	30.460	1.422
HUN	18.297	8.237	34.966	7.888	POL	28.915	25.955	31.695	1.498
SVK	19.045	7.159	33.516	8.856	BGR	29.427	26.585	32.565	1.626
PRT	23.291	13.150	37.918	7.032	GRC	32.510	28.605	38.225	3.055
GRC	23.789	14.294	32.506	5.794	ITA	33.450	28.443	37.609	3.012
CZE	24.151	11.893	44.296	9.526	ESP	35.071	28.188	40.780	3.624
SVN	24.247	11.256	42.431	8.675	SVK	37.432	32.610	41.544	2.800
ESP	27.509	14.830	43.496	8.522	SVN	39.166	31.935	45.553	2.942
ITA	31.445	20.101	45.723	7.246	IRL	40.220	34.543	47.597	4.208
FRA	31.945	18.915	50.993	9.124	GBR	40.706	31.964	45.930	4.932
GBR	32.638	17.736	49.932	9.369	DNK	41.813	33.062	55.056	5.783
FIN	33.291	17.229	53.172	10.604	AUT	46.527	43.224	50.507	2.208
BEL	34.969	20.223	56.349	10.690	CZE	46.731	42.985	51.262	2.357
DEU	35.540	21.669	57.530	11.091	DEU	47.448	43.305	50.304	2.018
SWE	36.753	20.718	56.632	11.132	FRA	47.515	40.750	52.237	3.908
DNK	36.950	19.777	62.090	12.240	NLD	62.782	56.554	68.026	3.479
AUT	37.485	21.241	60.418	11.767	FIN	65.481	54.423	75.459	5.639
NLD	39.087	20.750	61.285	12.005	SWE	67.211	58.691	77.932	5.828
IRL	42.179	15.098	89.684	20.751	BEL	68.011	59.094	74.007	4.658
LUX	73.451	33.755	124.591	27.945	LUX	90.402	71.375	114.632	13.103

Spain, and Portugal, TEC continued to increase mostly after 2016. Four of the five countries with the highest TEC, Finland, Sweden, Belgium, and the Netherlands, have kept it at a stable level over the long term. In Luxembourg, the country with the highest TEC, it decreased significantly in the period from 2005 to 2016.

Table 4 shows that the three countries Sweden, Austria, and Finland have a much higher average number of RES than the others. Figure 5 shows that these three countries have an almost stable trend in RES values throughout the study period. Spain, Portugal, and Slovenia also have a stable trend of RES since 2012. The countries with the highest increase in RES scores after the 2009 crisis are Denmark, Germany, Ireland, and the United Kingdom. Poland and Hungary have the lowest and very slowly increasing RES values.

Nuclear energy is generated in 14 of the 22 countries surveyed. Sweden and France have the highest values for the variable nuclear. As shown in Table 4, Finland and

Table 4 Descriptive statistics of variables RES and nuclear, 28 observations for every country, ordered according to mean

RES					Nuclear				
Country code	Mean	Min	Max	Std. dev	Country code	Mean	Min	Max	Std. dev
HUN	0.436	0.055	1.144	0.367	AUT	0	0	0	0
POL	0.607	0.112	1.637	0.557	DNK	0	0	0	0
CZE	1.187	0.376	2.297	0.725	GRC	0	0	0	0
NLD	1.240	0.187	3.238	0.872	IRL	0	0	0	0
BEL	1.307	0.139	4.056	1.326	ITA	0	0	0	0
GBR	1.319	0.272	4.375	1.273	LUX	0	0	0	0
LUX	1.346	0.506	3.113	0.671	POL	0	0	0	0
BGR	1.477	0.481	3.278	0.810	PRT	0	0	0	0
GRC	1.836	0.590	3.814	1.073	NLD	0.630	0.427	0.743	0.082
IRL	2.058	0.564	5.831	1.596	ROU	0.910	0.000	1.497	0.549
ROU	2.296	1.391	3.426	0.584	GBR	3.416	2.060	4.723	0.846
SVK	2.407	1.010	3.018	0.446	ESP	3.584	2.968	4.255	0.430
DEU	2.915	0.711	7.243	2.121	HUN	3.852	2.959	4.165	0.279
ITA	3.063	2.014	5.084	1.015	DEU	4.511	2.226	5.813	1.292
FRA	3.322	2.479	4.312	0.463	CZE	5.574	3.280	7.435	1.643
ESP	3.639	1.382	6.053	1.466	BGR	5.673	3.699	7.049	0.693
PRT	4.511	1.546	7.857	1.806	SVN	7.046	5.492	8.185	0.644
DNK	4.740	0.786	10.115	2.892	SVK	7.269	5.706	9.116	1.087
SVN	5.330	4.213	8.141	0.850	BEL	10.972	5.850	13.277	1.930
FIN	11.952	9.583	14.063	1.414	FIN	11.262	9.846	12.462	0.697
AUT	13.840	11.765	15.515	0.926	FRA	18.022	15.186	20.025	1.461
SWE	22.874	16.933	26.539	2.299	SWE	19.299	14.598	23.385	2.819

Belgium still reach double-digit average values. Figure 6 shows that France, Sweden, Finland, Germany, and Belgium have a long-term tendency to decrease the values of the variable nuclear, with two of them, Sweden and Belgium, showing strong fluctuations from the trend. The conflict in Ukraine in 2022 is expected to cause structural breaks in the downward trends of the nuclear variable. In other countries (see Fig. 6), the use of nuclear power plants has been almost stable since 2010.

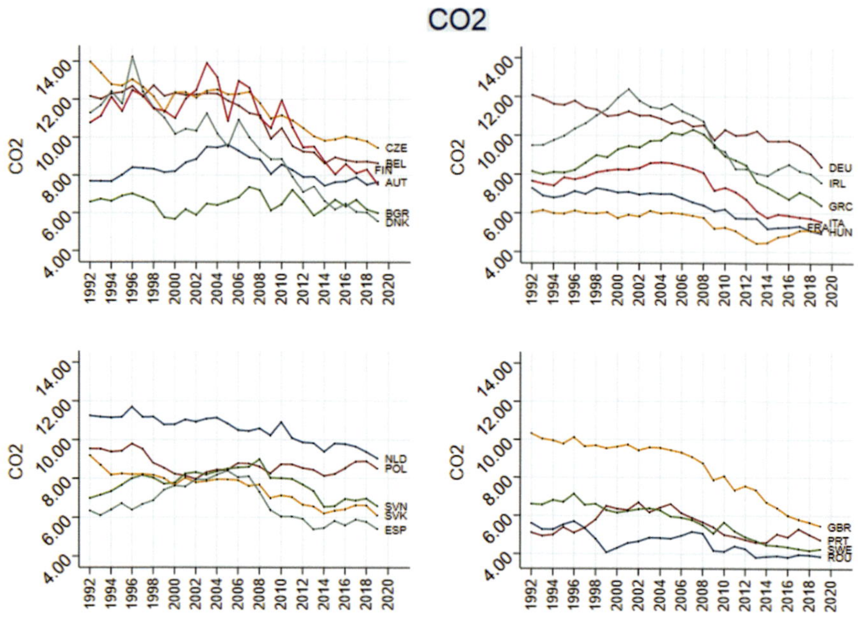

Fig. 1 CO_2 emissions per capita in tons, 21 counties except LUX

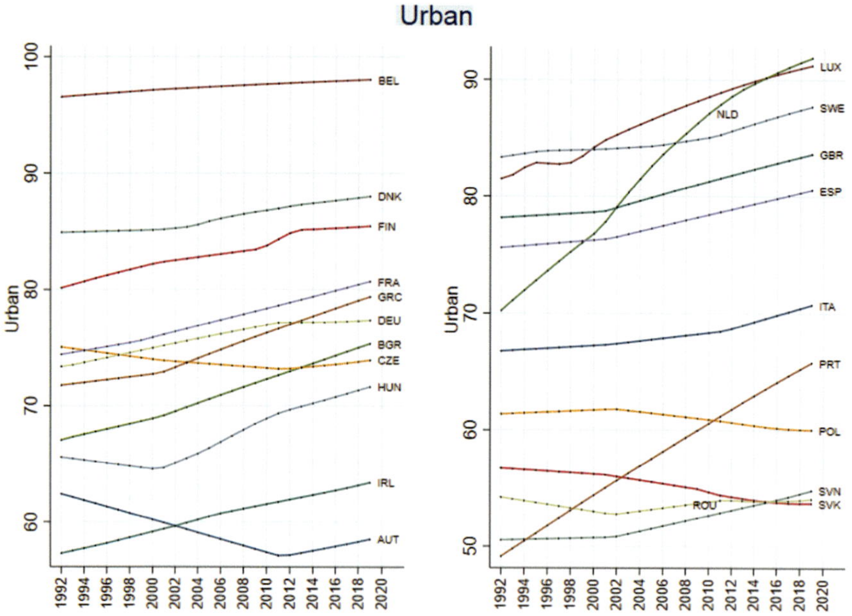

Fig. 2 Urban population (% of total population), 22 countries

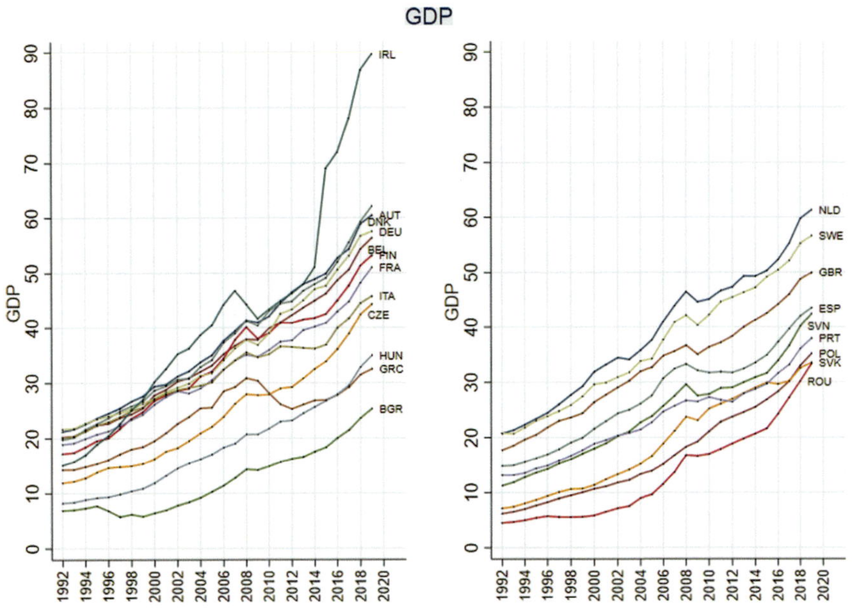

Fig. 3 GDP per capita in thousands USD, 21 countries except LUX

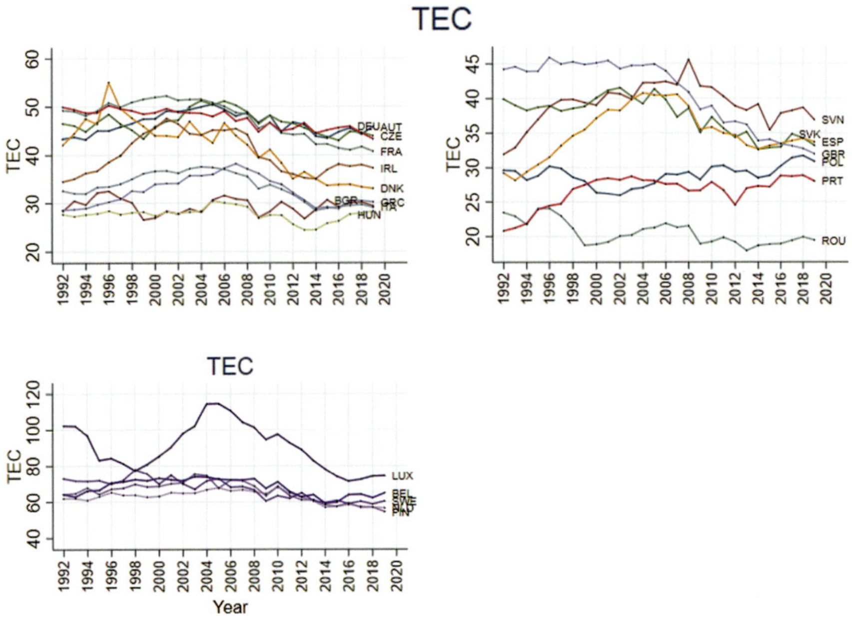

Fig. 4 Total energy consumption per capita in MWh (TEC), 22 countries

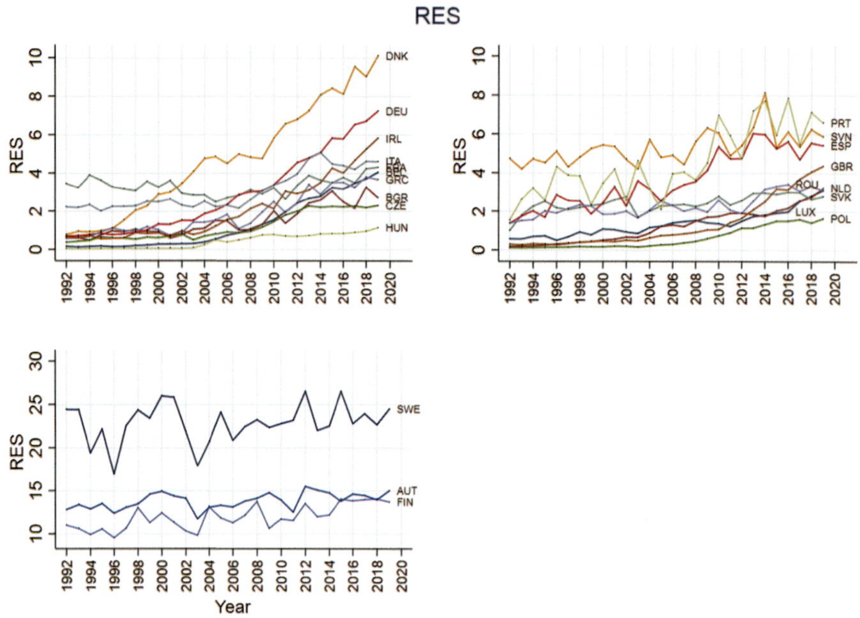

Fig. 5 Energy produced from renewable sources per capita in MWh (RES), 22 countries

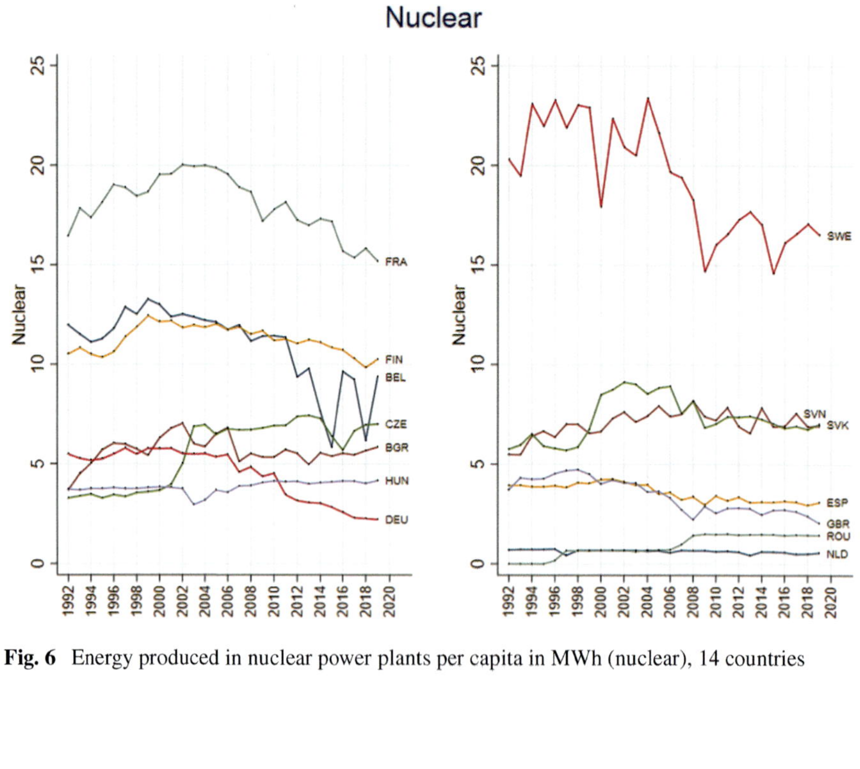

Fig. 6 Energy produced in nuclear power plants per capita in MWh (nuclear), 14 countries

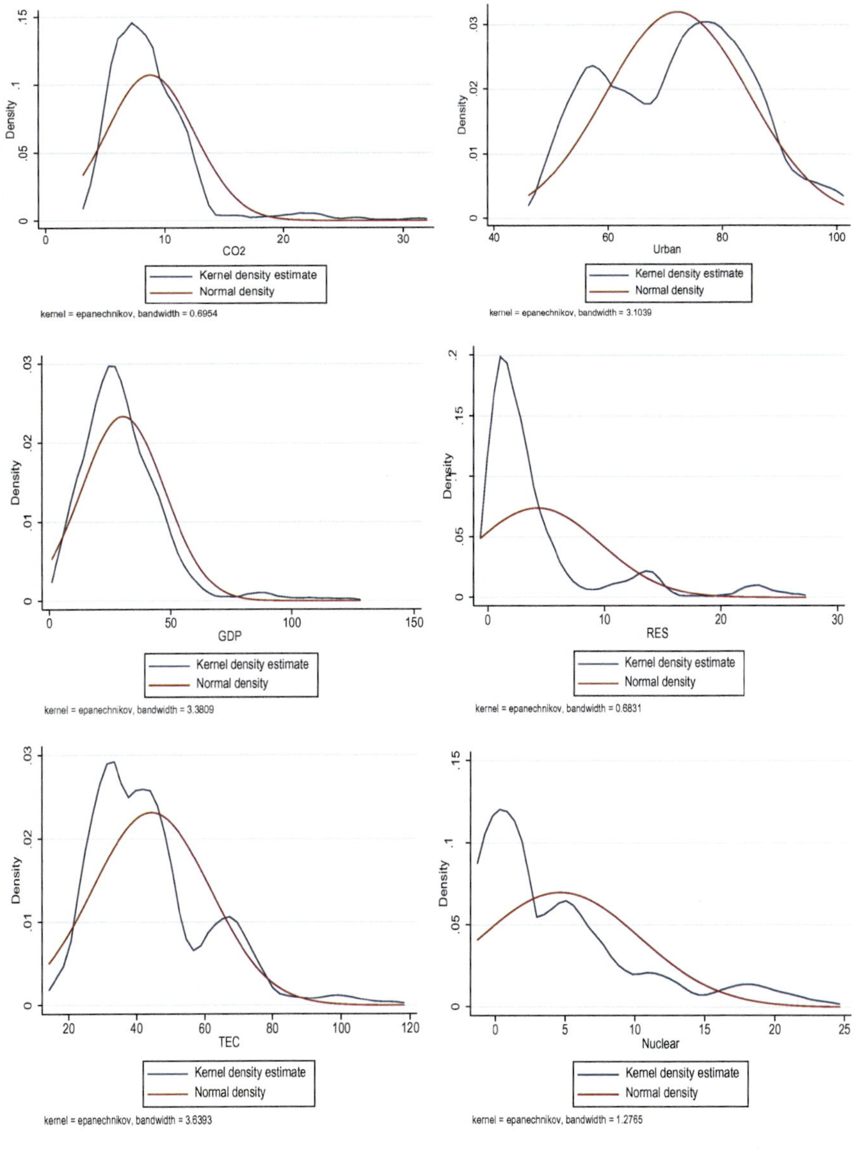

Fig. 7 Kernel density estimation. *Source* Own processing in Stata

3.2 *Overview of the Methods for Investigating the Relationship Between Variables*

Various methods were used to investigate the relationship between the variables under study. A progression from simpler to more complex methods was chosen.

Significant results from the simpler method whose diagnostics led to doubts about the validity of its application were used to select parameters in the more complex method subsequently used. The chapter concludes with the resulting model describing the nexus between the dependent variable Log_CO_2 and the independent variables Log_ Urban, Log_GDP, Log_ RES, Log_ TEC, and Nuclear. The list of used methods is as follows:

- preliminary analysis,
- multiple regression,
- Threshold model: cross-sectional model with threshold variable (urbanization size of country),
- Panel model (Fixed Effects (FE), Random Effects (RE), Least Squares Dummy Variables (LSDV),
- Panel threshold models with threshold variable (urbanization rate of the country),
- Fully Modified OLS (FMOLS),
- Dynamic OLS (DOLS),
- Panel VAR model (General moments methods—GMM, impulse response functions—IRF),
- Pooled Mean Group Estimation (PMG),
- Mean Group Estimation (MG), and
- Dynamic FE estimation.

Threshold models (first proposed by Tong (2012)) are characterized by linear processes that are divided into parts by values of a threshold variable. Due to their rich dynamic structure, these models have the ability to capture nonlinearities and asymmetries that occur in data from different domains (Gonzalo and Pitarakis 2002; Hansen 1997). The panel analysis detailed in Baltagi (2005) allows models with one or more levels of fixed effects or random effects to be applied to multilevel longitudinal data. Panel data are measurements of the variable under consideration for the same set of N cases (entities, individuals, etc.) at multiple time points T, allowing the identification and control of individual effects and dynamics. A limitation of the linear model is the assumption that a monotonic change in the predictor variable leads to a monotonic change in the dependent variable. If we assume a change in model behavior after a threshold value of a variable is crossed, then linear panel regression is not sufficient. A suitable tool then is the panel threshold regression model with individual-specific effects proposed by Caner and Hansen (2004) and Hansen (1999, 2011), which combines the two approaches. The basic idea of nonlinear threshold panel data models is to find out whether the marginal effect of the explanatory variables on the explained variable changes significantly before or after the threshold value of the threshold variable (Yanlong 2018).

Ahmad et al. (2020), in their review article, provide an insight into the evolution of methodologies used in nexus studies in general, which they divide into seven generations of methodological breakthroughs. The sixth generation of methods, which are used later in this chapter consists of panel analysis, including PVECM, FMOLS, and DOLS. This generation is associated with studies that apply advanced panel data techniques, including unit root tests and cointegration relationships within a panel data set. Foundational papers in this class include Pedroni (1998) and Pesaran et al. (1999). The basic idea of these studies is that a very long-run relationship is likely to become common across countries and is therefore estimated by pooling all the data. This in turn can be considered as a generalized nexus within a panel of countries (Pesaran et al. 1999). The advantage of panel data is that there is an opportunity to test for country-specific nexus along with panel nexus analysis. At the beginning of the analysis, the data needs to be tested for stationarity and panel cointegration. The suitable methodology apparatus should be employed in order to capture the panel nexus, namely the Panel Vector Error-Correction Model (PVECM) (Hansen 1999; Menegaki 2014); Fully Modified Ordinary Least Square (FMOLS) (Pedroni 2001); Generalized Method of Moments (GMM) (Bond and Windmeijer 2002); or Dynamic OLS (DOLS) (Mark and Sul 2003).

4 Preliminary Analysis

The preliminary analysis will serve to justify why it is necessary to work with transformed variables in further analyses. Figure 7 gives an overview of the empirical probability of the measurements in the set of all 616 measurements for each variable obtained by Kernel density estimation. Kernel density estimates for individual variables are multimodal, asymmetric, and have long tails. To overcome these shortcomings, log-transformed data are included in the further analyses.

The following correlation coefficients in Table 5 were calculated from the original, untransformed data and serve to provide initial brief information about the relationship between the variables under study.

The correlation coefficients for the variable pairs TEC with CO_2 and Urban with TEC reach the highest values. All correlation coefficients are positive except for the variable pairs RES with CO_2 and Nuclear with CO_2. However, this fact is not surprising. One of the interesting points in this part is to find out the relationship between the Urban variable and others. The scatter plots for the Urban variable and other variables are shown in Fig. 8. Here it can be seen visually that the values scatter

Table 5 Correlation coefficients from untransformed data

	CO_2	Urban	GDP	RES	TEC	Nuclear
CO_2	1					
Urban	0.3841	1				
GDP	0.3848	0.4553	1			
RES	−0.2240	0.1214	0.2230	1		
TEC	0.7540	0.6464	0.5231	0.278	1	
Nuclear	−0.1679	0.3261	−0.0550	0.4620	0.3551	1

Source Own processing in Stata

the most in the case of the Urban pair with RES. The last figure also shows that many European countries do not generate nuclear electricity.

As mentioned earlier, the transformed data were included in the analyses. Figure 9 was produced after all variables with non-zero values were transformed with the decimal logarithm and Kernel density estimation was applied. The Kernel density estimation no longer has long one-sided tails, but its shape does not indicate the normality of the distribution.

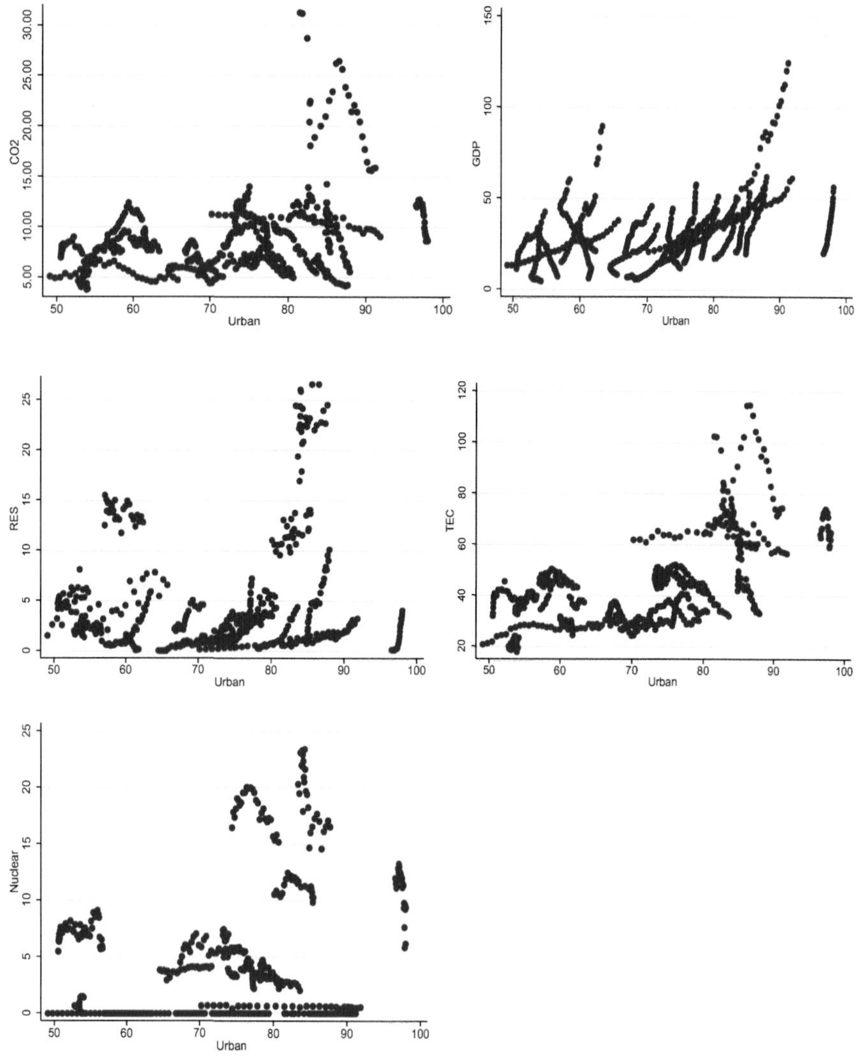

Fig. 8 Scatterplots of the variable Urban with other variables. *Source* Own processing in Stata

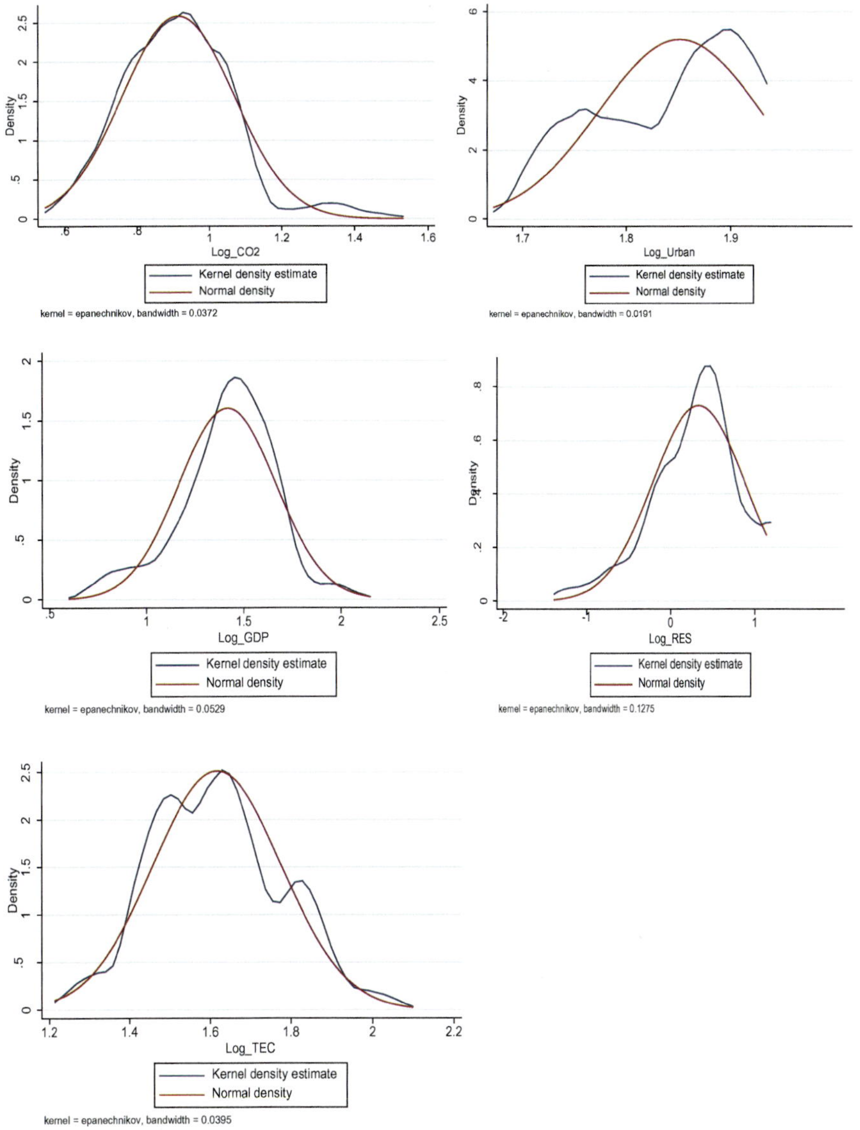

Fig. 9 Kernel density estimation after logarithmic transformation. *Source* Own processing in Stata

Table 6 Multiple regression (Pooled model)

Log_CO_2	Coef	Std. Err	t	P > |t|	95% Conf. interval		Beta
Log_Urban	−0.1878	0.0447	−4.19	0.000	−0.2757	−0.0998	−0.0936
Log_GDP	−0.0758	0.0148	−5.13	0.000	−0.1049	−0.0468	−0.1224
Log_RES	−0.0714	0.0057	−12.65	0.000	−0.0825	−0.0603	−0.2534
Log_TEC	1.0495	0.0229	45.77	0.000	1.0045	1.0946	1.0812
Nuclear	−0.0130	0.0005	−24.19	0.000	−0.0140	−0.0119	−0.4831
Cons	−0.2438	0.0670	−3.63	0.000	−0.3756	−0.1121	

Source Own processing in Stata

5 Results: Model Development

5.1 Multiple Regression

In the previous section, a simple correlation analysis of the input variables was presented. The following two models (multiple regression and threshold model) will be constructed without considering the time variable. In this section, multiple regression of the transformed variables is performed, considering Log_CO_2 as the dependent variable and Log_Urban, Log_GDP, Log_ RES, Log_TEC, and Nuclear as independent variables. The following results were recorded: F test (p-value) = 0.000, R-squared = 0.85. The resulting values and tests for the significance of the regression coefficients are shown in Table 6.

All variables are significant in the model and all but one, Log_ TEC, have a negative impact on CO_2 emissions. A comparison of the beta coefficients shows that Log_Urban has the least influence on Log_CO_2 and Log_ TEC has the strongest influence of the variables considered. The positive effect of the use of nuclear energy and renewable energy sources on the reduction of CO_2 emissions confirms the result obtained in Table 6.

5.2 Threshold Models

The previous result of multiple linear regression is a simplified approach to complex problems. Multiple regression does not provide a complete picture of the true form of the relationships under investigation because, for example, it does not take into account jumps or structural breaks. Another model not considering the time variable is the threshold model. Threshold models (TAR) have been created to describe jumps or structural breaks in the relationships between variables. According to Tsay (2010), the TAR model uses piecewise linear models to get a better approximation to the conditional means equation. It uses threshold space to improve the linear approximation. It is necessary to select a threshold variable and regional variables in the model. Regional variables are those for which a change in the relationship with the dependent variable is expected after the threshold value for the threshold variable is exceeded. For a threshold variable, the optimization task is to find one or more thresholds for the selected model.

The first threshold models considered are models in which Log_Urban is both a threshold and a regional variable. The other variables were gradually selected as dependent variables. The threshold effect only applies to Log_CO_2 and Log_TEC (Table 7). The effect of urbanization on emissions after exceeding 1.790 (Log Urban) has weakened and the effect of urbanization on total energy consumption after exceeding 1.802 (Log Urban) has strengthened.

The next step is to consider a model in which Log_Urban is the threshold variable, Log_CO_2 is the dependent variable, and the regional variables are all other variables (Table 8). The model calculated a threshold value for Log_Urban of 1.920 (83% Urban). Table 8 shows that the threshold effect only applies to the variable Log_GDP. The effect of GDP on CO_2 emissions is negative in both regions (regimes). In the first regime, when urbanization is less than 83%, Log_CO_2 decreases with a coefficient of 0.0847 when Log_GDP increases. When urbanization is higher than 83%, Log_CO_2 decreases even faster (0.1907) as Log_GDP increases. Eight countries have urbanization higher than 83%. These countries are BEL, DNK, FIN, LUX, NLD, SWE, and GB. This means that GDP growth has a stronger impact on the decrease of CO_2 emissions when the urbanization level is above 83%.

Table 7 Two threshold models with dependent Log_CO_2 and Log_TEC

Dependent: Log_CO_2		Log_Urban threshold: 1.790			
		Coef	P > |z|	95% Conf. Int	
Region 1	Log Urban	2.342	0.00	1.598	3.086
	Cons	−3.238	0.00	−4.540	−1.935
Region 2	Log Urban	1.277	0.00	1.006	1.549
	Cons	−1.482	0.00	−1.996	−0.968
Dependent: Log_TEC		*Log_Urban threshold: 1.802*			
		Coef	*P > |z|*	*95% Conf. Int*	
Region 1	Log Urban	1.073	0.000	0.521	1.624
	Cons	−0.362	0.463	−1.329	0.604
Region 2	Log Urban	2.421	0.000	2.181	2.661
	Cons	−2.927	0.000	−3.382	−2.472

Source Own processing in Stata

Table 8 Threshold model—complex

Dependent: Log_CO_2		Log_Urban threshold: 1.920			
		Coef	P > |z|	95% confident interval	
Region1	**Log_GDP**	−0.0847	0.000	−0.1114	−0.0581
	Log_RES	−0.0522	0.000	−0.0632	−0.0411
	Log_TEC	1.0620	0.000	1.0174	1.1065
	Nuclear	−0.0106	0.000	−0.0118	−0.0095
	cons	−0.6020	0.000	−0.6624	−0.5417
Region2	**Log_GDP**	−0.1907	0.000	−0.2690	−0.1123
	Log_RES	−0.0629	0.000	−0.0857	−0.0401
	Log_TEC	1.1317	0.000	1.0186	1.2449
	Nuclear	−0.0180	0.000	−0.0201	−0.0159
	cons	−0.5606	0.000	−0.7176	−0.4036

Source Own processing in Stata

5.3 *Panel Models*

In addition to the cross-sectional variables considered so far, the panel model includes time variables. Thus, 22 countries (N = 22) are analyzed, the period from 1992 to 2019 (T = 28), a total of 616 measurements. The appropriate model is selected from: Fixed Effects (FE), Least Squares Dummy Variable Model (LSDV), Random Effects (RE), and pooled model. The tests and parameters of the models FE and RE are given in Table 9. The results of the FE test (F test) are favorable, p-value = 0.00, we also reject the hypothesis that all specific constants are equal to zero (F test), p-value = 0.00. The results of the RE test are also favorable (Wald test), p-value = 0.00. The regression coefficients of the models FE and RE are close to each other and have the same signs as the Pooled model in Table 6. The coefficient on Nuclear is significant only for RE. The increase in Urban, GDP, RES, and, in the case of RE, Nuclear, leads to a decrease in CO_2 emissions per capita. In contrast, an increase in TEC leads to an increase in per capita emissions. Thus, according to the panel models, higher urbanization has the effect of decreasing CO_2 emissions.

The Hausman test and the Breusch-Pagan Lagrange multiplication test (Table 10) were used to test the correctness of model selection (FE, RE, Pooled). The results of these two tests suggest that a fixed effects model should be preferred.

Individual countries can thus be distinguished on the basis of specific constant μi. Specific constants can be determined using the Least Squares Dummy Variable Model (LSDV, Table 11). AUT was chosen as the reference country, which corresponds to the absolute term −0.2978. The country-specific constants are the sum of −0.2978 and Coef from Table 11. It is clear that this is very heterogeneous data, as the differences are significant with 4 exceptions (FIN, HUN, SVN, SVK). The biggest positive differences to the reference country (AUT) are in CZE, DNK, GRC, IRL, and POL. On the other hand, the largest negative differences are found in the countries BEL, FRA, NLD, and SWE.

Table 9 Fixed effects and random effects model

Dependent: Log_CO_2	Fixed effects model		Random effects model	
	Coef	P > |t|	Coef	P > |t|
Log_Urban	−0.3261	0.000	−0.3547	0.000
Log_GDP	−0.0774	0.000	−0.0744	0.000
Log_RES	−0.0404	0.000	−0.0430	0.000
Log_TEC	1.1752	0.000	1.1698	0.000
Nuclear	−0.0013	0.136	−0.0028	0.001
Cons	−0.2534	0.044	−0.1882	0.117

Source Own processing in Stata

Table 10 Model fit tests

Test	H_0	p
Hausman	Preferred model is RE	0.001
Breusch-Pagan Lagrange multiplier	No panel effect	0.000
Pesaran CD test	Residuals are not correlated	0.000
Modified Wald test for groupwise heteroskedasticity	Homoskedasticity (or constant variance)	0.000

Source Own processing in Stata

Table 11 Least squares dummy variable model

Country	Coef	p-value	Country	Coef	p-value
AUT	−0.2978	0.013	ITA	0.1041	0.000
BEL	−0.0450	0.026	LUX	0.1084	0.000
BGR	0.0800	0.000	NLD	0.0478	0.000
CZE	0.1192	0.000	POL	0.1779	0.000
DNK	0.1336	0.000	SVN	0.0163	0.078
FIN	−0.0041	0.784	SVK	0.0110	0.245
FRA	−0.0935	0.000	ESP	0.0589	0.000
DEU	0.1036	0.000	ROU	0.0770	0.000
GRC	0.1766	0.000	PRT	0.0650	0.000
HUN	0.0174	0.193	SWE	−0.2694	0.000
IRL	0.1165	0.000	GBR	0.0705	0.000

Source Own processing in Stata

However, the results of the Pesaran CD test and the Modified Wald test for groupwise heteroscedasticity in Table 10 show that the residuals of the selected Fixed Effects model do not fulfill the conditions of non-correlation and homoscedasticity. The chosen model therefore has reserves in terms of suitability for the given data.

5.4 Panel Threshold Model

Since the Fixed Effects model shows signs of cross-correlation and heteroscedasticity, the threshold model is used in the following analysis. The variable Log_Urban is chosen as the threshold and all other variables were considered as regional variables in the first stage. However, in the search for a suitable model, it was found that the confidence intervals of the coefficients of the variable Nuclear from the two regimes overlapped. Therefore, the variable Nuclear was ultimately not considered as a regional variable. In the following model, the dependent variable is Log_CO_2, the independent variable is Nuclear, the regional variables are Log_GDP, Log_RES, and Log_TEC, and the threshold variable is Log_Urban. At this stage, a choice is made between models with one or two thresholds for the variable Log_Urban. Assuming that the model has two thresholds, their estimation is shown in Table 12.

Two thresholds are estimated for Log_Urban: 1.7568 and 1.8519. The corresponding values for the Urban variable are 57.12 and 71.10%. The F test with bootstrap gives a statistic of 136.17 ($p = 0.047$) for the single threshold and 76.13 ($p = 0.133$) for the double threshold. Since the single option has a p-value of less than 0.005 and the double option has a p-value of more than 0.05, the resulting model is selected as the single threshold. The threshold estimated with this model is 1.87519 (71.10%). Urbanization and the parameters of this model are shown in Table 13.

The F test of the overall significance of the model and the F test that all residuals $ui = 0$ also show p-values = 0.000. The effect of the variable nuclear is similar to

Table 12 Threshold estimator (significancy level = 0.95)

Model	Threshold	Lower bound	Upper bound
Threshold—1	1.8519	1.8502	1.8530
Threshold—21	1.8519	1.8508	1.8530
Threshold—22	1.7568	1.7544	1.7570

Source Own processing in Stata

Table 13 Panel threshold model—threshold value Log_Urban = 1.8519

Dependent Log_CO_2		Coef	p	95% Confident Interval	
Nuclear		−0.0023	0.004	−0.0040	−0.0007
Log_GDP	Region 1	−0.0632	0.000	−0.0787	−0.0476
	Region 2	−0.1655	0.000	−0.1869	−0.1440
Log_RES	Region 1	−0.0524	0.000	−0.0636	−0.0411
	Region 2	−0.0269	0.000	−0.0381	−0.0158
Log_TEC	Region 1	1.0759	0.000	1.0296	1.1221
	Region 2	1.1874	0.000	1.1428	1.2321
Cons		−0.7331	0.000	−0.8077	−0.6586

Own processing—in Stata

the previous case (Table 9). For the other variables, a threshold effect is applied in terms of the urbanization of the country. When the urbanization threshold of 71% is exceeded:

- the impact of GDP on the reduction of CO_2 emissions increases,
- the impact of renewables on the reduction of CO_2 emissions weakens, and
- the impact of total energy consumption on the growth of CO_2 emissions increases.

Previous models were developed without taking into account the stationary nature of the individual variables and their mutual cointegration. In the following text, the given variables are examined to see if they fulfill these properties.

5.5 *Panel Unit Root Tests*

Unit root tests are used to determine the type of a random variable, i.e., whether the variable is a non-stationary process of type I (1) (Choi 2001; Harris and Tzavalis 1999; Kwiatkowski et al. 1992). First, the logarithmic variable itself was tested and then its first difference. The following tests were used to test the unit root:

- Fisher (Augmented Dickey-Fuller Test)-Time trend, Lagged difference 1,
- Fisher (Phillips-Perron unit root test)-Time trend, Lagged difference 1,
- Im-Pesaran-Shin (Im et al. 2003)-Time trend, Lagged specification 1,
- Levin-Lin-Chu (Levin et al. 2002)-Time trend, Lagged specification 1,
- Breitung-Time trend, Lagged difference 1, and
- Hardri-Time trend.

Rejection of the null hypothesis in all tests except the Hardri test (Hadri 2000) means stationarity. In case of Hadri test, acceptance of the null hypothesis means stationarity in all panels. The test results (p-value) are shown in Table 14. The table shows that the variables on the level (except Log_Urban and Log_RES) have a unit root, but their difference is already stationary. This means that the variables are first-order integrated—I (1). The variable Log_Urban is stationary, except in the Breitung (Breitung and Das 2005) and Hadri tests, and its first difference has a unit root. In the case of the variable Log_RES, tests that consider the variable Log_RES to be stationary predominate. The only test that considers Log_RES as I (1) is the Hadri test. Results that deviate from most tests are shown in grey in the table.

5.6 *Panel Cointegration Tests*

The development of economic time series, whose correlation is justified by some economic theories, does not diverge over time. After short-term fluctuations, the

Table 14 Panel unit root tests

Variables	Panel unit root test (p-value)						
	Fisher (ADF)		Fisher (PP)	IPS	LLC	Breitung	Hadri
Log_Urban	P	0.000	0.0059	0.0040	0.000	0.1427	0.000
	Z	0.0012	0.0126				
	L*	0.0005	0.0013				
	Pm	0.000	0.0019				
D.Log_Urban	P	0.2940	0.9999	0.5613	0.0012	0.0101	0.000
	Z	0.6293	0.9998				
	L*	0.4973	0.9995				
	Pm	0.3131	0.9981				
Log_GDP	P	0.8839	0.9977	0.5506	0.010	0.0783	0.000
	Z	0.5765	0.9883				
	L*	0.5878	0.9827				
	Pm	0.8761	0.9905				
D.Log_GDP	P	0.000	0.000	0.000	0.000	0.000	0.000
	Z	0.000	0.000				
	L*	0.000	0.000				
	Pm	0.000	0.000				
Log_CO_2	P	0.9171	0.9339	0.8553	0.0202	0.9543	0.000
	Z	0.9002	0.9398				
	L*	0.8967	0.9391				
	Pm	0.9052	0.9113				
D.Log_CO_2	P	0.000	0.000	0.000	0.000	0.000	0.4880
	Z	0.000	0.000				
	L*	0.000	0.000				
	Pm	0.000	0.000				
Log_TEC	P	0.7400	0.5265	0.5297	0.0009	0.7906	0.000
	Z	0.5714	0.7502				
	L*	0.5537	0.6720				
	Pm	0.7516	0.5543				
D.Log_TEC	P	0.000	0.000	0.000	0.000	0.000	0.1404
	Z	0.000	0.000				
	L*	0.000	0.000				
	Pm	0.000	0.000				
Log_RES	P	0.0001	0.000	0.003	0.0329	0.003	0.000
	Z	0.0013	0.000				
	L*	0.0005	0.000				
	Pm	0.000	0.000				

(continued)

Table 14 (continued)

Variables	Panel unit root test (p-value)			IPS	LLC	Breitung	Hadri
	Fisher (ADF)		Fisher (PP)				
D.Log_RES	P	0.000	0.000	0.000	0.000	0.000	0.9453
	Z	0.000	0.000				
	L*	0.000	0.000				
	Pm	0.000	0.000				
Nuclear	P	0.8234	0.0996	*	0.9138	0.2019	0.000
	Z	0.2295	0.0133				
	L*	0.1999	0.004				
	Pm	0.8240	0.0932				
D.Nuclear	P	0.000	0.000	*	0.0002	0.000	0.9977
	Z	0.000	0.000				
	L*	0.000	0.000				
	Pm	0.000	0.000				

Source Own processing in Stata

system returns to equilibrium. Statisticians refer to this state as cointegration. Cointegration tests are used to determine whether there is a long-term stable relationship between variables. Table 15 shows the results of three types of cointegration tests:

- KAO test—Lags (1),
- Pedroni test—AR parameter is panel-specific, includes panel-specific time trend, Lags (1), and

Table 15 Panel cointegration tests; H0: No cointegration, default H1: All panels are cointegrated

		Panel Cointegration test	p-value
Kao test	1	Modified Dickey-Fuller	0.0302
	2	Dickey-Fuller	0.0312
	3	Augmented Dickey-Fuller	0.4347
	4	Unadjusted modified Dickey-Fuller	0.0000
	5	Unadjusted Dickey-Fuller	0.0006
Pedroni test	1	Modified Philips-Perron	0.0014
	2	Philips-Perron	0.0001
	3	Augmented Dickey-Fuller	0.0001
Westerlund test	1	Group-mean variance ratio variance, H1: All panels are cointegrated	0.0412
	2	Group-mean variance ratio variance, H1: Some panels are cointegrated	0.4315

Source Own processing in Stata

- Westerlund tests—include panel-specific time trends.

Of the ten tests, two do not reject the null hypothesis of no cointegration (the third Kao test and the second Westerlund test). We can therefore proceed with the estimation of the cointegration relationship between the variables analyzed.

5.7 *Long-Run and Short-Run Estimation*

Since least squares estimation is not consistent in panel data, fully modified OLS (FMOLS) and dynamic OLS (DOLS) are used to estimate long-run cointegration relationships in panel data. They were developed by Kao and Chiang (2000) and Pedroni (2000). DOLS uses past (lag) and future values of variable differences (leads) and thus takes into account the presence of autocorrelation and endogeneity of variables. The "xtcointreg" command (Stata 15.1) was used to estimate cointegration vectors. The estimates of the cointegration coefficients with FMOLS and for the whole set also for DOLS are captured in Table 16.

The estimated coefficients for the EU-22 can be calculated as the average of the estimated coefficients for the individual countries (the so-called mean group). Most of the regression coefficients are significant. The coefficients for Log_TEC (with one exception—NLD) are significant and all are positive, which means that an increase in energy consumption leads to an increase in CO_2 emissions. The least significant coefficients are in the case of the variable Nuclear. The explanation for this phenomenon is the fact that many European countries do not operate nuclear power plants. The signs of the regression coefficients are relatively uniform for each variable. The exception is the variable Log_Urban, where the signs of the coefficients are different for each country. This means that urbanization has a positive impact on CO_2 emissions but in some countries this impact is negative. The coefficient in Log_Urban for EU-22, which is calculated as the average of all coefficients, has a negative value (-0.30). This means that CO_2 emissions in the EU-22 as a whole decrease with increasing urbanization. For comparison, the last row of the table shows the coefficients of the DOLS model for the EU22. The coefficients are different for the Log_Urban and Log_GDP variables than in the FMOLS model. Compared to the FMOLS model, urbanization has a larger effect on CO_2 emissions reduction and the effect of GDP on emissions reduction is the opposite than in the FMOLS case.

5.7.1 Panel VAR Estimation Results

In the FD specification, we use the second lags of the untransformed variables as the earliest lag used as an instrument—instlags(2/5) first difference (FD) (Andrews and Lu 2001). The results of estimating the first-order panel model VAR according to the GMM (Amisano and Giannini 1997; Bond and Windmeijer 2002) pattern and in FD are presented in Table 17.

Table 16 Estimates of cointegration coefficients using FMOLS

	Panel FMOLS				
	Log_Urban	Log_GDP	Log_RES	Log_TEC	Nuclear
AUT	−0.49***	−0.04***	−0.48***	1.35***	0.00
BEL	−13.20	−0.10	−0.02	1.25***	−0.01***
BGR	2.35***	−0.09*	−0.10	1.16***	−0.02***
CZ	0.28	−0.07 ***	−0.05***	1.04***	−0.01***
DNK	3.41	−0.43***	0.05	1.49***	−0.01***
FIN	−0.08	0.01	−0.32***	1.82***	−0.01***
FRA	1.68****	−0.20***	−0.12***	1.72***	−0.01***
DEU	4.17***	−0.56***	0.02**	1.44***	−0.02***
GRC	−0.60	−0.18*	−0.01	1.48***	0.00
HUN	0.15**	−0.10***	−0.02***	1.12***	−0.03***
IRL	0.62**	−0.09***	−0.07***	1.10***	0.00
ITA	−3.40***	0.10***	−0.12***	0.90***	0.00
LUX	6.20***	−0.76 ***	0.02	1.02 ***	0.00
NLD	0.25***	−0.07***	−0.05***	0.60	−0.00
POL	−1.50***	−0.09***	−0.02***	0.96***	0.00
SVN	−0.53	−0.02	−0.21***	1.31***	−0.01***
SVK	1.89***	0.02	−0.13***	0.90***	−0.01***
ESP	0.19	−0.13	−0.19***	1.27***	−0.01**
ROU	−2.56***	-0.01	−0.18***	1.31***	0.02***
PRT	−1.27***	0.21***	−0.21***	1.21***	0.00
SWE	−3.80***	−0.11***	−0.48***	1.32***	−0.01***
GBR	−0.34	−0.17***	−0.03	1.36***	−0.03***
EU-22	**−0.30*** **	**−0.13*** **	**−0.12*** **	**1.23*** **	**−0.01**
	Panel DOLS				
EU-22	**−4.02*** **	**0.01*** **	**−0.07*** **	**0.94*** **	**−0.01**

Source Own processing in Stata

In contrast to the long-run relationship, this relationship is not so strong in the case of the short run. Of the 30 regression coefficients, only 7 are significant. For the dependent variable Log_Urban, only the variable Log_GDP has a significant coefficient. This means that GDP growth in year t-1 could lead to an increase in urbanization in year t. The independent variable Log_Urban is not significant for any of the other five variables at the position of the dependent variable. It can be said that an increase in TEC in year t-1 could lead to a decrease in both GDP and RES in year t. Other highlighted cases can be commented similarly.

Table 17 Result of first-order panel VAR model estimation in the GMM FD framework

Equation	Log_CO_2		Log_Urban		Log_GDP	
	Coef	P_{value}	Coef	P_{value}	Coef	P_{value}
$L1Log\_CO_2$	0.9873	0.004	−0.0047	0.123	0.2621	0.154
L1Log_Urban	0.2914	0.696	0.9655	0.000	0.2694	0.476
L1Log_GDP	−0.1475	0.066	0,0021	**0.001**	0.9405	0.000
L1Log_RES	0.0554	0.339	−0.0004	0.383	0.0412	0.210
L1Log_TEC	−0.1267	0.808	0.0039	0.345	−0.7725	**0.019**
L1Nuclear	−0.0079	0.118	−9.93–E6	0.815	0.0003	0.932
Equation	*Log_RES*		*Log_TEC*		*Nuclear*	
	Coef	P_{value}	*Coef*	P_{value}	*Coef*	P_{value}
$L1Log\_CO_2$	1.6630	0.077	0.7353	**0.003**	44.27	**0.038**
L1Log_Urban	−0.8402	0.792	0.2993	0.592	25.09	0.610
L1Log_GDP	0.6240	**0.006**	0.0048	0.931	1.2871	0.793
L1Log_RES	0.6673	0.000	0.0358	0.394	1.1702	0.714
L1Log_TEC	−3.1868	**0.032**	−0.1330	0.713	−49.65	0.123
L1Nuclear	0.0128	0.360	−0.0074	**0.040**	0.4298	0.228

Source Own processing in Stata

5.7.2 Panel VAR-Granger Causality

Another test performed was the Granger causality test (Granger 1969). The hypotheses for the panel VAR-Granger causality Wald test are as follows:

Ho: Excluded variable does not Granger-cause Equation variable.

Ha: Excluded variable Granger-causes Equation variable.

The individual Equation variables and Excluded variables are evident from Table 18. The same results were obtained as in Table 17, since the first-order VAR model was used.

5.7.3 Eigenvalue Stability Condition

Lütkepohl (2005) shows that a VAR model is stable if all moduli of the eigenvalues of the companion matrix are strictly less than 1. Stability implies that the panel VAR is invertible and has an infinite-order vector moving-average (VMA) representation, providing known interpretation to estimated impulse response functions (IRFs) (Abrigo and Love 2016). All eigenvalues lie within the unit circle (Table 19). The PVAR satisfies the stability condition.

Table 18 Result of panel VAR-Granger causality Wald test

Equation/Excluded	p_{value}				
Log_CO_2		Log_Urban		Log_GDP	
Log_Urban	0.696	Log_CO_2	0.123	Log_CO_2	0.154
Log_GDP	0.066	Log_GDP	0.001	Log_Urban	0.476
Log_RES	0.339	Log_RES	0.383	Log_RES	0.210
Log_TEC	0.808	Log_TEC	0.345	Log_TEC	0.019
Nuclear	0.118	Nuclear	0.815	Nuclear	0.932
All	0.159	All	0.003	All	0.069
Log_RES		Log_TEC		Nuclear	
Log_CO_2	0.077	Log_CO_2	0.003	Log_CO_2	0.038
Log_Urban	0.792	Log_Urban	0.592	Log_Urban	0.610
Log_GDP	0.006	Log_GDP	0.931	Log_GDP	0.793
Log_TEC	0.032	Log_RES	0.394	Log_RES	0.714
Nuclear	0.360	Nuclear	0.040	Log_TEC	0.123
All	0.003	All	0.000	All	0.171

Source Own processing in Stata

Table 19 Roots of the companion matrix

Real	Imaginary	Modulus
0.99377	0	0.99377
0.91228	0	0.91228
0.83841	0	0.83841
0.60392	−0.10245	0.61255
0.60392	0.10245	0.61255
−0.09468	0	−0.09468

Source Own processing in Stata

5.7.4 Impulse Response Function

Impulse responses for all variables were also analyzed. Almost all impulse responses in this model can be considered as zero. Figure 10 shows plots for the Log_CO_2 response, which can also be considered zero.

The exception is the option shown in Fig. 11. The response to the TEC impulse, reflected in the GDP variable, can be seen as a non-zero response. However, this response is lost over the course of the year. This result is consistent with the results in Table 17.

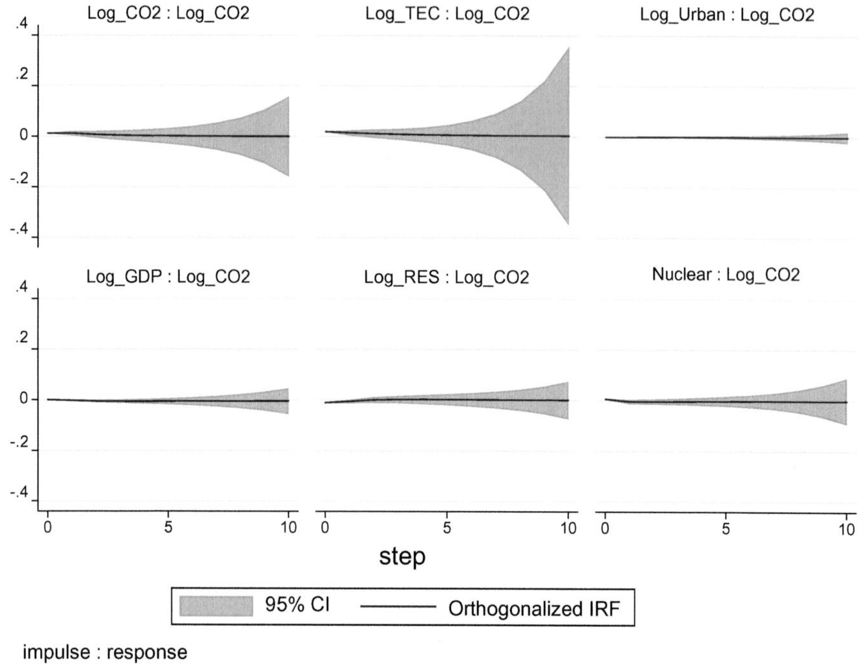

Fig. 10 Impulse response function (response CO_2). *Source* Own processing in Stata

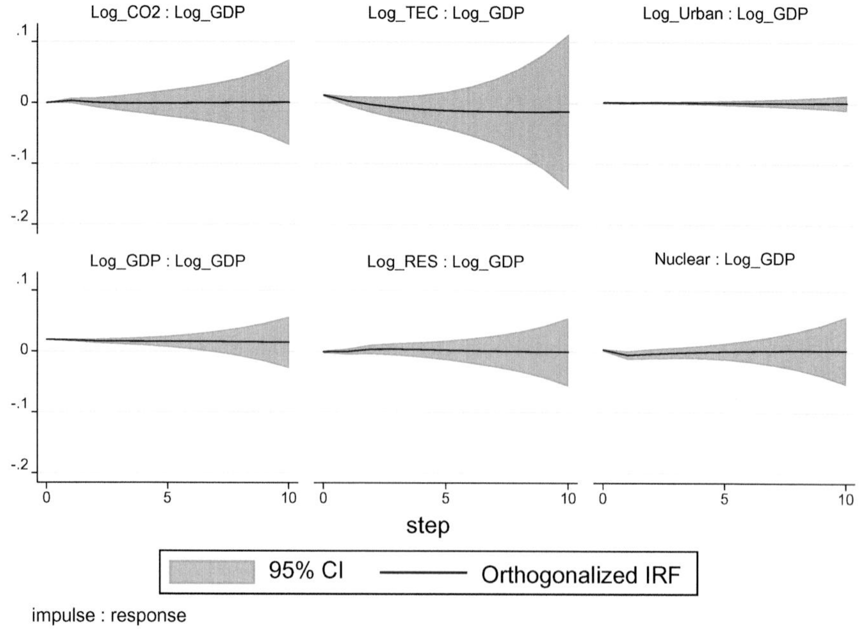

Fig. 11 Impulse response function (response GDP). *Source* Own processing in Stata

5.7.5 Estimation of Heterogeneous Panels

The approaches used to estimate the cointegration relationship are DOLS (Dynamic OLS), FMOLS (Fully Modified OLS), MG (Mean Group Estimator), and PMG (Pooled Mean Group Estimator). Pesaran et al. (2001) proposed MG and PMG estimators that account for both short-run and long-run dynamics. They introduced a restriction on the coefficients of the long-run relationship, which can be the same for all entities. They used the maximum likelihood method to estimate the PMG model. This regression can also be estimated with different coefficients of the long-run relationship for each entity, from which the average is then calculated, giving the MG estimator.

5.7.6 Pooled Mean Group Estimation

The dynamic PMG estimator constrains the cointegration vector coefficients to be the same for all entities. The speed of adjustment and the short-run coefficients may be different for each entity. In this model (Table 20), the coefficient on the Urban variable is significant, as are the other coefficients corresponding to the long-run relationship at the top of the table (ec). In terms of the short-run relationship (SR) at the bottom of the table, the coefficients on Log_GDP and Log_Urban are not significant.

Table 20 Pooled mean group estimation, error-correction form (PMG)

D.Log_CO_2		Coef	p_{value}	95% Conf. interval	
ec	Log_Urban	−1.9124	0.000	−2.3520	−1.4728
	Log_GDP	−0.1016	0.000	−0.1270	−0.0762
	Log_RES	−0.0928	0.000	−0.1171	−0.0685
	Log_TEC	1.1426	0.000	1.0440	1.2412
	Nuclear	−0.0097	0.000	−0.0123	−0.0071
SR	ec	−0.2079	0.000	−0.2912	−0.1245
	D.Log_Urban	−0.5383	0.852	−6.1752	5.0985
	D.Log_GDP	0.0295	0.356	−0.0330	0.0920
	D.Log_RES	−0.0959	0.000	−0.1447	−0.0469
	D.Log_TEC	0.8927	0.000	0.7321	1.0531
	D.Nuclear	−0.0074	0.000	−0.0112	−0.0035
	Cons	0.5849	0.000	0.3510	0.8187

Source Own processing in Stata

Table 21 Mean group estimation: error-correction form (MG)

D.Log_CO_2		Coef	p_{value}	95% Conf. interval	
ec	Log_Urban	−1.8208	0.199	−4.6015	0.9599
	Log_GDP	−0.1194	0.241	−0.3192	0.0803
	Log_RES	−0.1304	0.003	−0.2178	−0.0431
	Log_TEC	0.9309	0.000	0.6816	1.1803
	Nuclear	0.0001	0.987	−0.0126	0.0128
SR	ec	−0.6219	0.000	−0.7653	−0.4784
	D.Log_Urban	−17.6639	0.335	−53.5640	18.2361
	D.Log_GDP	0.0506	0.238	−0.0335	0.1348
	D.Log_RES	−0.0196	0.325	−0.0587	0.0194
	D.Log_TEC	0.4824	0.000	0.2828	0.6819
	D.Nuclear	−0.0034	0.141	−0.0080	0.0011
	Cons	2.7268	0.138	−0.8803	6.3340

Source Own processing in Stata

5.7.7 Mean Group Estimation

In the case of Mean Group Estimation, the coefficients for both the long-run and short-run relationship are calculated separately for each entity and then the mean is calculated from them (Table 21). Compared to PMG, there are fewer significant coefficients, but the signs for the coefficients are the same.

5.7.8 Dynamic FE Estimation

The Dynamic FE estimator limits the coefficients of the cointegration vector, the speed of adjustment, and the short-run coefficients so that they are the same for all entities. When using this model, the significant coefficient in the long-term relationship is only for the variable GDP and TEC (Table 22).

Overall, in all three models, the coefficients have the same signs except for the variable Nuclear in the long-run relationship. However, the significance of the coefficient changes across models. The PMG model has the most significant coefficients. Of particular interest are the error-correction speed of adjustment parameter (ec), and long-run coefficients. If ec is significantly negative, then there is a long-run relationship between the dependent variable Log_CO_2 and the independent variables Log_Urban, Log_GDP, Log_RES, Log_TEC, and Nuclear, and these variables show a return to long-run equilibrium. The speed of adjustment shows the following values −0.20788 for PMG, −0.6219 for MG, and −0.1535 for DFE. In all three cases, p-value = 0. The Hausman test can be used to select a suitable model. In this test, the null hypothesis is of the form:

H0: Difference in coefficients is not systematic.

Table 22 Dynamic fixed effects regression (DFE)

D.Log_CO_2		Coef	p_{value}	95% Conf. interval	
ec	Log_Urban	−0.4165	0.120	−0.9417	0.1086
	Log_GDP	−0.1042	0.001	−0.1634	−0.0449
	Log_RES	−0.0311	0.079	−0.0659	0.0035
	Log_TEC	1.3522	0.000	1.1614	1.5429
	Nuclear	0.0001	0.981	−0.0067	0.0068
SR	ec	−0.1535	0.000	−0.2016	−0.1055
	D.Log_Urban	−0.1006	0.874	−1.3479	1.1466
	D.Log_GDP	0.0231	0.023	−0.0412	0.0875
	D.Log_RES	−0.0925	0.000	−0.1058	−0.0791
	D.Log_TEC	0.8719	0.000	0.7904	0.9534
	D.Nuclear	−0.0068	0.000	−0.0085	−0.0051
	Cons	−0.0542	0.479	−0.2043	0.0959

Source Own processing in Stata

The results of the tests are presented in Table 23.

For the first test, it is a choice between the MG and the PMG model. The result of the first test is that the PMG estimator is preferred over MG. For the second test, the choice is between MG and the DFE test. The result of the second test is that the MG model is preferred over the DFE model. This means that the PMG model will be preferred in the next analysis. Its full version is in Table 24.

The first part of the table is common to all countries studied (identical to Table 20) and reflects the cointegration (long-run relationship between variables). The short-run relationship between the variables (first differences), which is specific to each country, is described in the remaining part of the table. For the long-run relationship, all coefficients are significant and negative except Log_TEC.

This means that in the long run, an increase in Urban, GDP, RES, and Nuclear causes a decrease in CO_2 emissions. The increase in TEC causes an increase in CO_2 emissions. The error-correction rate of adjustment parameters (ec) is significantly negative in 15 cases (at 0.1 significance level). In the short run, there are much fewer significant coefficients. The coefficients of the variable Log_TEC show the strongest significance, which are consistently positive and significant except in two cases (SVK, PRT). The coefficients for the variable Log_Urban show the significance (at the 0.1 significance level) in the smallest number of countries, precisely in 8 cases.

Table 23 Hausman tests

	Test	Models	p_{value}
1	Hausman	MG, PMG	0.550
2	Hausman	MG, DFE	0.000

Source Own processing in Stata

Table 24 PMG full model

D.Log_CO$_2$	LR	Coef	p_{value}			
ec	Log_Urban	−1.9124	0.000			
	Log_GDP	−0.1016	0.000			
	Log_RES	−0.0928	0.000			
	Log_TEC	1.1426	0.000			
	Nuclear	−0.0097	0.000			
D.Log_CO$_2$	**SR**	Coef	p_{value}	**SR**	Coef	p_{value}
AUT	ec	−0.0607	0.505	BEL	−0.3622	0.005
	D.Log_Urban	−0.8922	0.284		−50.4271	0.233
	D.Log_GDP	0.0252	0.863		−0.0143	0.926
	D.Log_RES	−0.3927	0.000		0.0224	0.343
	D.Log_TEC	1.1958	0.000		0.6348	0.000
	D.Nuclear	0			−0.0045	0.001
	Cons	0.1602	0.507		1.0913	0.014
BGR	ec	−0.0496	0.051	CZ	−0.3405	0.002
	D.Log_Urban	6.6001	0.381		7.3482	0.071
	D.Log_GDP	−0.0953	0.101		−0.0051	0.960
	D.Log_RES	−0.0893	0.000		−0.0530	0.011
	D.Log_TEC	1.2524	0.000		0.6465	0.000
	D.Nuclear	−0.0215	0.000		−0.0116	0.000
	Cons	0.1325	0.045		0.9954	0.006
DNK	ec	−0.3859	0.039	FIN	−0.1078	0.244
	D.Log_Urban	12.4111	0.218		1.4594	0.634
	D.Log_GDP	−0.2348	0.192		0.0628	0.718
	D.Log_RES	−0.0926	0.100		−0.2689	0.000
	D.Log_TEC	0.7492	0.001		1.5841	0.000
	D.Nuclear	0			−0.0142	0.023
	Cons	1.1650	0.045		0.3198	0.250
FRA	ec	−0.3283	0.000	DEU	−0.0008	0.981
	D.Log_Urban	16.3687	0.003		5.2581	0.037
	D.Log_GDP	−0.0434	0.499		0.1311	0.137
	D.Log_RES	−0.0874	0.000		−0.0138	0.589
	D.Log_TEC	1.2950	0.000		1.1547	0.000
	D.Nuclear	−0.0144	0.000		−0.0090	0.005
	Cons	0.9225	0.000		−0.0077	0.944
GRC	ec	−0.2478	0.000	HUN	−0.01208	0.620
	D.Log_Urban	7.6734	0.042		0.2636	0.811
	D.Log_GDP	−0.0579	0.555		−0.2346	0.049

(continued)

Table 24 (continued)

D.Log_CO_2	LR	Coef	p_{value}			
	D.Log_RES	−0.0322	0.043		−0.0129	0.315
	D.Log_TEC	0.5347	0.002		1.0342	0.000
	D.Nuclear	0			−0.0294	0.000
	Cons	0.7161	0.000		0.0357	0.586
IRL	ec	−0.0317	0.377	ITA	−0.2431	0.084
	D.Log_Urban	1.7447	0.797		2.6309	0.441
	D.Log_GDP	0.0034	0.952		0.0366	0.694
	D.Log_RES	−0.0484	0.004		−0.0738	0.008
	D.Log_TEC	0.9771	0.000		0.7558	0.000
	D.Nuclear	0			0	
	Cons	0 0.0812	0.439		0.6823	0.082
LUX	ec	0.1035	0.015	NLD	−0.0148	0.535
	D.Log_Urban	3.1559	0.065		−1.2500	0.456
	D.Log_GDP	0.0537	0.577		−0.1501	0.275
	D.Log_RES	−0.0442	0.038		−0.0392	0.346
	D.Log_TEC	1.3445	0.000		0.9427	0.000
	D.Nuclear	0			−0.0049	0.760
	Cons	−0.3209	0.012		0.0492	0.500
POL	ec	−0.0899	0.096	SVN	−0.2598	0.063
	D.Log_Urban	−6.1839	0.067		6.7630	0.109
	D.Log_GDP	0.0531	0.723		0.1124	0.463
	D.Log_RES	−0.0255	0.292		−0.1476	0.000
	D.Log_TEC	0.7348	0.000		0.8890	0.000
	D.Nuclear	0			−0.0130	0.001
	Cons	0.2445	0.114		0.6688	0.063
SVK	ec	−0.7456	0.000	ESP	−0.5253	0.000
	D.Log_Urban	−2.3666	0.386		4.0326	0.540
	D.Log_GDP	0.2038	0.031		0.1974	0.233
	D.Log_RES	0.0069	0.771		−0.0670	0.008
	D.Log_TEC	0.0543	0.756		0.6629	0.004
	D.Nuclear	−0.0062	0.022		−0.0196	0.013
	Cons	1.9599	0.000		1.5152	0.002
ROU	ec	−0.1697	0.070	PRT	−0.3697	0.000
	D.Log_Urban	0.4682	0.879		−19.6583	0.035
	D.Log_GDP	0.0668	0.509		0.3216	0.064
	D.Log_RES	−0.1433	0.000		−0.0997	0.000
	D.Log_TEC	0.9591	0.000		0.0209	0.906

(continued)

Table 24 (continued)

D.Log_CO_2	LR	Coef	p_{value}			
	D.Nuclear	0.0087	0.557		0	
	Cons	0.4414	0.072		1.0768	0.001
SWE	ec	−0.2122	0.024	GBR	−0.1184	0.004
	D.Log_Urban	−11.8561	0.044		4.6129	0.125
	D.Log_GDP	0.3185	0.055		−0.1031	0.508
	D.Log_RES	−0.4063	0.000		0.0003	0.990
	D.Log_TEC	1.1448	0.000		1.0703	0.000
	D.Nuclear	−0.0059	0.020		−0.0164	0.000
	Cons	0.6008	0.038		0.3369	0.008

Source Own processing in Stata

The positive value of the regression coefficient was recorded in CZ, FRA, DEU, GRC, LUX, and the negative value in POL, PRT, and SWE. To illustrate, we present the PMG model for France:

$$D.Log\_CO_{2t} = -0.3283\begin{pmatrix} Log\_CO_{2t-1} - 1.9124\,Log\_Urban_t - 0.1016 \\ Log\_GDP_t - 0.0928\,Log\_RES_t + 1.1426 \\ D.Log\_TEC_t - 0.0097\,Nuclear_t \end{pmatrix}$$
$$+\,16.3687\,D.Log\_Urban_t - 0.0874\,D.Log\_RES_t$$
$$+\,1.2950\,D.Log\_TEC_t - 0.0144\,D.Nuclear_t + 0.9225$$

Comparing this model with the model for SVK, it turns out (not shown here) that the SVK short-run relationship has fewer significant coefficients and the coefficient on D.Log_GDP has a positive effect.

6 Conclusions

The aim of this chapter was to examine the impact of urbanization level, GDP per capita, total energy consumption per capita, consumption of energy produced from renewable sources per capita, and energy consumption from nuclear power plants per capita on the level of CO_2 emissions per capita. For this purpose, the following estimates were used:

- multiple regression,
- Threshold model: cross-sectional model with threshold variable (level of urbanization of the country),
- Panel model (Fixed Effects (FE), Random Effects (RE), Least Squares Dummy Variables (LSDV)),

- Panel threshold models with threshold variable (level of urbanization of the country),
- Fully Modified OLS (FMOLS),
- Dynamic OLS (DOLS),
- Panel VAR model (GMM, IRF),
- Pooled Mean Group Estimation (PMG),
- Mean Group Estimation (MG), and
- Dynamic FE estimation.

Using multiple regression, it was shown that all coefficients in the regression equation are significant and all variables except total energy consumption have a negative effect on the CO_2 emission level. This means that as they increase, the amount of emissions decreases. The mitigation effect of renewables and nuclear power on CO2 emissions is well known and is confirmed by a number of studies (Azam et al. 2021; Cheng et al. 2019; Gozgor et al. 2018; Nathaniel et al. 2021; Nathaniel and Iheonu 2019; Salari et al. 2021; Vural 2020; Wang et al. 2021). Moreover, many of these studies confirm also positive effect of renewables on economic growth (Cheng and Liu 2019). In addition to economic benefits and reducing greenhouse gas emissions and thus mitigating climate change, the wider use of renewable energy sources is associated with a number of other social, economic, and societal benefits, such as new job opportunities (Fragkos and Paroussos 2018; Proença and Fortes 2020), alleviation of energy poverty (Zhao et al. 2022), reduction of the consumption of non-renewable resources and preserving them for future generations, as well as increasing energy security and reducing energy dependence on imports of energy raw materials from third countries (Mathews and Tan 2014; Sachs et al. 2019). The countries of the European Union should therefore develop and support the implementation of policies, strategies, and projects that increase the use of renewable energy sources. These benefits could also be used as a counterargument to advocate the need to decarbonize the European economy and meet the objectives of the European Green Deal, which are threatened by the economic losses caused by the COVID-19 pandemic (Kahraman et al. 2022) and the war conflict in Ukraine. Comparison of the beta coefficients shows that urbanization (Log_Urban) has the weakest effect on Log_ CO_2 of the variables considered and total energy consumption (Log_TEC) has the strongest effect.

For the threshold models, we first considered the cases where the threshold variable and the regional variable are Log_Urban and the other variables are dependent variables. The threshold effect only applied to Log_CO_2 and Log_TEC, so that the effect of urbanization on CO_2 emissions became weaker after exceeding the threshold of 1.790 (61.65%—Urban), and the effect of urbanization on total energy consumption became stronger after exceeding the threshold of 1.802 (63.38%—Urban). Although urbanization has only a modest impact on the growth of CO_2 emissions, the countries with the higher rate of urbanization in particular have a huge potential for promoting the use of renewable energy sources and increasing energy efficiency. Urban development policies therefore should be focused on promoting resilience of the cities, through energy management of buildings, insulation, smart grid applications, green infrastructure, etc.

Applying a more complex threshold model, where Log_Urban was the threshold variable, Log_CO_2 was the dependent variable and the rest of the variables were the regional variables, showed that the threshold effect only applied to the variable Log_GDP. The model calculated a threshold value for Log_Urban of 1.920 (83% Urban). The effect of GDP on CO_2 emissions is negative in both regions (regimes). In the first regime, when urbanization is less than 83%, Log_CO_2 decreases as Log_GDP increases (with a coefficient of −0.0847). When urbanization is higher than 83%, Log_CO_2 decreases even faster with increasing Log_GDP (−0.1907).

According to all three-panel models (FE, RE, Pooled), higher urbanization rates have an impact on the decrease in CO_2 emissions. Testing the selected fixed effects model (FE) showed that its residuals did not meet the conditions of uncorrelatedness and homoscedasticity, so the panel threshold model was applied in the next section. The test showed that the model with a threshold of 1.87519 (71.10% urbanization) was the best fit. It was found that when the threshold of 71% urbanization is exceeded,:

- the impact of GDP on the decline of CO_2 emissions is amplified,
- the impact of renewables on the decline in CO_2 emissions is weakened, and
- the impact of total energy consumption on CO_2 emissions growth is amplified.

The tests have shown that the level variables (except Log_Urban and Log_RES) have a unit root, but their difference is already stationary. That is, the variables are integrated of first order—I(1). The variable Log_Urban, with the exception of the Breitung test, is stationary and its first difference has a unit root. In the case of RES, tests that treat the Log_RES variable as stationary dominate. As for cointegration, two of the ten tests used reject the null hypothesis of no cointegration.

FMOLS and DOLS procedures were used to estimate the long-run cointegration relationships (long-run estimation) in the panel data. A comparison of the signs of the regression coefficients in FMOLS across countries shows considerable agreement. The exception is the Log_Urban variable, where the signs of the coefficients differ across countries. This means that urbanization has a positive effect on CO_2 emissions in some countries but a negative effect on CO_2 emissions in some countries. The coefficient for Log_Urban for EU22, which is calculated as the average of all coefficients, has a negative value (−0.30). This means that overall, in the EU-22 CO_2 emissions decrease with increasing urbanization. Comparing the results obtained by FMOLS and DOLS for the EU-22, they are found to be different for the variables Log_Urban and Log_GDP. Urbanization has a stronger effect on reducing emissions according to DOLS (−4.02) than according to FMOLS (−0.30) and GDP growth leads to a decrease (−0.13) in CO_2 emissions according to FMOLS and an increase (0.01) in CO_2 emissions according to DOLS.

A VAR model was used to estimate the short-run estimation. In contrast to the long-run relationship, the relationship is not so strong in the short-run case. Of the 30 regression coefficients, only 6 are significant. These conclusions are confirmed by Granger causality tests and the results of Impulse Response Functions analyses. The following models were used to estimate non-stationary heterogeneous panels: Pooled

Mean Group Estimation (PMG), Mean Group Estimation (MG), and Dynamic Fixed Effects Regression (DFE). These models take into account both short-run and long-run dynamics. When comparing the coefficients in these models, they were found to have the same signs of the coefficients. The Hausman test to select the appropriate model led to the application of the PMG model, which has a common cointegration part for all countries. In this model, all coefficients corresponding to the long-run (cointegrating) relationship are significant from the EU-22 perspective and all are negative except Log_TEC. This means that in the long run, growth in urbanization, GDP, renewable energy consumption, and nuclear energy consumption contribute to the reduction of CO_2 emissions. On the other hand, growth in total energy consumption contributes to an increase in CO_2 emissions. This model confirms the findings of the previous simpler model, underlying that the interdependence between variables is present in the long run. It is therefore important to design and implement energy and climate as well as urban development policies and strategies that will enable the implementation of measures that will bear fruit in the long term.

In the short term (EU-22), the pattern is not so clear. The coefficients for urbanization and GDP are not significant. Regarding the effects of the other coefficients, we can say that if the increase in renewables increases, the decrease in emissions also increases, and the same is true for energy produced by nuclear power plants. Since the model works with logarithmic differences, it is possible to say that when the increase in energy consumption increases, the increase in CO_2 emissions also increases. Using the PMG—Full model, a short run part was also obtained for each country, leading to the finding that there are much less significant coefficients in the short run. The coefficients for the variable Log_TEC, which are still positive and significant in all but two cases, show the greatest significance and uniformity. The coefficients for the variable Log_Urban show the least significance (at the 0.1 significance level). Finally, in order to achieve the ambitious goals of carbon neutrality and balanced development, EU countries need to implement energy and climate policies that take into account energy security and environmental protection, while at the same time not compromising the economic development of the region. Economic development should go hand in hand with the promotion of sustainable urban planning that will ensure a healthy and resilient environment for their inhabitants.

Acknowledgements This work was supported by the Scientific Grant Agency of the Ministry of Education, Science, Research and Sport of the Slovak Republic and the Slovak Academy of Sciences under grant VEGA 1/0508/21.

References

Abdel-Rahman AN, Safarzadeh MR, Bottomley MB (2006) Economic growth and urbanization: a cross-section and time-series analysis of thirty-five developing countries. Rivista Internazionale Di Scienze Economiche e Commerciali 53(3):334–348. https://doi.org/10.1007/BF03029785

Abrigo MRM, Love I (2016) Estimation of panel vector autoression in Stata. Stand Genomic Sci 16(3):778–804

Adebayo TS, Rjoub H, Akinsola GD, Oladipupo SD (2022) The asymmetric effects of renewable energy consumption and trade openness on carbon emissions in Sweden: new evidence from quantile-on-quantile regression approach. Environ Sci Pollut Res 29(2):1875–1886. https://doi.org/10.1007/S11356-021-15706-4/FIGURES/4

Ahmad N, Aghdam RF, Butt I, Naveed A (2020) Citation-based systematic literature review of energy-growth nexus: An overview of the field and content analysis of the top 50 influential papers. Energy Econ 86:104642

Akbostanci E, Türüt-Aşik S, Tunç GI (2009) The relationship between income and environment in Turkey: is there an environmental Kuznets curve? Energy Policy. https://doi.org/10.1016/j.enpol.2008.09.088

Alam MJ, Begum IA, Buysse J, Rahman S, Van Huylenbroeck G (2011) Dynamic modeling of causal relationship between energy consumption, CO2 emissions and economic growth in India. Renew Sustain Energy Rev 15(6):3243–3251

Ali R, Bakhsh K, Yasin MA (2019) Impact of urbanization on CO2 emissions in emerging economy: evidence from Pakistan. Sustain Cities Soc 48:101553. https://doi.org/10.1016/J.SCS.2019.101553

Al-Mulali U, Ozturk I (2016) The investigation of environmental Kuznets curve hypothesis in the advanced economies: the role of energy prices. Renew Sustain Energy Rev. https://doi.org/10.1016/j.rser.2015.10.131

Al-mulali U, Che Sab CNB (2018) Energy consumption, CO2emissions, and development in the UAE. Energy Sour Part B: Econ Plan Policy. https://doi.org/10.1080/15567249.2012.689796

Amin SB, Alam T (2018) The relationship between energy consumption and sectoral output in Bangladesh: an empirical analysis. J Dev Areas 52(3):39–54

Amisano G, Giannini C (1997) Topics in structural VAR econometrics, 2nd edn. Springer

Andrews DWK, Lu B (2001) Consistent model and moment selection procedures for GMM estimation with application to dynamic panel data models. J Econ 101:123–164

Ang JB (2007) CO2 emissions, energy consumption, and output in France. Energy Policy. https://doi.org/10.1016/j.enpol.2007.03.032

Anwar A, Sarwar S, Amin W, Arshed N (2019) Agricultural practices and quality of environment: evidence for global perspective. Environ Sci Pollut Res 26(15):15617–15630. https://doi.org/10.1007/S11356-019-04957-X/TABLES/9

Apergis N, Christou C, Gupta R (2017) Are there environmental kuznets curves for US state-level CO2 emissions? Renew Sustain Energy Rev. https://doi.org/10.1016/j.rser.2016.11.219

Apergis N, Payne JE (2009) CO2 emissions, energy usage, and output in Central America. Energy Policy 37(8):3282–3286

Apergis N, Ozturk I (2015) Testing environmental Kuznets curve hypothesis in Asian countries. Ecol Ind. https://doi.org/10.1016/j.ecolind.2014.11.026

Armeanu DS, Joldes CC, Gherghina SC, Andrei JV (2021) Understanding the multidimensional linkages among renewable energy, pollution, economic growth and urbanization in contemporary economies: quantitative assessments across different income countries' groups. Renew Sustain Energy Rev 142:110818. https://doi.org/10.1016/J.RSER.2021.110818

Arminen H, Menegaki AN (2019) Corruption, climate and the energy-environment-growth nexus. Energy Econ 80:621–634. https://doi.org/10.1016/J.ENECO.2019.02.009

Arrow K, Bolin B, Costanza R, Dasgupta P, Folke C, Holling CS, Pimentel D (1995) Economic growth, carrying capacity, and the environment. Ecol Econ 15(2):91–95

Arrow K, Bolin B, Costanza R, Dasgupta P, Folke C, Holling CS (1996) Economic growth, carrying capacity, and the environment. Environ Dev Econ. https://doi.org/10.1017/S1355770X00000413

Ayres RU, van den Bergh JCJM (2005) A theory of economic growth with material/energy resources and dematerialization: interaction of three growth mechanisms. Ecol Econ 55(1):96–118. https://doi.org/10.1016/j.ecolecon.2004.07.023

Azam A, Rafiq M, Shafique M, Zhang H, Yuan J (2021) Analyzing the effect of natural gas, nuclear energy and renewable energy on GDP and carbon emissions: a multi-variate panel data analysis. Energy 219:119592. https://doi.org/10.1016/J.ENERGY.2020.119592
Bai X, Chen J, Shi P (2012) Landscape urbanization and economic growth in China: positive feedbacks and sustainability dilemmas. Environ Sci Technol 46(1):132–139. https://doi.org/10.1021/ES202329F/SUPPL_FILE/ES202329F_SI_001.PDF
Bakirtas T, Akpolat AG (2018) The relationship between energy consumption, urbanization, and economic growth in new emerging-market countries. Energy. https://doi.org/10.1016/j.energy.2018.01.011
Balcilar M, Ozdemir ZA, Yetkiner H (2014) Are there really bubbles in oil prices? Phys A: Stat Mech Appl. https://doi.org/10.1016/j.physa.2014.09.020
Baltagi BH (2005) Econometric analysis of panel data. John Wiley & Sons
Bekhet HA, Othman NS (2018) The role of renewable energy to validate dynamic interaction between CO2 emissions and GDP toward sustainable development in Malaysia. Energy Econ. https://doi.org/10.1016/j.eneco.2018.03.028
Bilgili F, Koçak E, Bulut Ü (2016) The dynamic impact of renewable energy consumption on CO2 emissions: a revisited environmental Kuznets curve approach. Renew Sustain Energy Rev. https://doi.org/10.1016/j.rser.2015.10.080
Bo S (2011) A literature survey on environmental Kuznets curve. Energy Procedia. https://doi.org/10.1016/j.egypro.2011.03.229
Bölük G, Mert M (2014) Fossil & renewable energy consumption, GHGs (greenhouse gases) and economic growth: evidence from a panel of EU (European Union) countries. Energy. https://doi.org/10.1016/j.energy.2014.07.008
Bond SR, Windmeijer F (2002) Finite sample inference for GMM estimators in linear panel data models.
Breitung J, Das S (2005) Panel unit root tests under cross-sectional dependence. Stat Neerl 59:414–433
Brook BW, Alonso A, Meneley DA, Misak J, Blees T, van Erp JB (2014) Why nuclear energy is sustainable and has to be part of the energy mix. Sustain Mater Technol 1–2:8–16. https://doi.org/10.1016/J.SUSMAT.2014.11.001
Buongiorno J, Corradini M, Parsons J, Petti D (2019) Nuclear energy in a carbon-constrained world: Big challenges and big opportunities. IEEE Power Energy Mag 17(2):69–77. https://doi.org/10.1109/MPE.2018.2885250
Cai Y, Sam CY, Chang T (2018) Nexus between clean energy consumption, economic growth and CO2 emissions. J Clean Prod 182:1001–1011. https://doi.org/10.1016/J.JCLEPRO.2018.02.035
Caner M, Hansen BE (2004) Instrumental variable estimation of a threshold model. Econ Theory 20:813–843
Chen PY, Chen ST, Hsu CS, Chen CC (2016) Modeling the global relationships among economic growth, energy consumption and CO2 emissions. Renew Sustain Energy Rev 65:420–431. https://doi.org/10.1016/J.RSER.2016.06.074
Cheng M, Liu B (2019) Analysis on the influence of China's energy consumption on economic growth. Sustainability 11(14):3982. https://doi.org/10.3390/SU11143982
Cheng C, Ren X, Wang Z, Yan C (2019) Heterogeneous impacts of renewable energy and environmental patents on CO2 emission - evidence from the BRIICS. Sci Total Environ 668:1328–1338. https://doi.org/10.1016/J.SCITOTENV.2019.02.063
Choi I (2001) Unit root tests for panel data. J Int Money Financ 20:249–272
Cirella GT (2020) Sustainable human-nature relations: environmental scholarship: economic evaluation, urban strategies. Springer, Berlin. https://doi.org/10.1007/978-981-15-3049-4
Cirella GT (2022) Human settlements: urbanization, smart sector development, and future outlook. Springer, Berlin. https://doi.org/10.1007/978-981-16-4031-5
Cirella GT, Tao L (2008) Measuring sustainability: an application using the index of sustainable functionality in South East Queensland, Australia. Int J Interdiscip Soc Sci: Annu Rev 3. https://doi.org/10.18848/1833-1882/CGP/v03i08/52680

Cirella GT, Tao L (2009a) An adaptive quantitative method to measure sustainability: an application for the State of Queensland, Australia. Int J Environ Cult Econ Soc Sustain Annu Rev 5:127–139. https://doi.org/10.18848/1832-2077/CGP/v05i01/54563
Cirella GT, Tao L (2009b) The index of sustainable functionality: an application for measuring sustainability. World Acad Sci Eng Technol 3:268–274. https://doi.org/10.5281/zenodo.1330369
Cirella GT, Zerbe S (2014a) Index of sustainable functionality: application in Urat front banner. In: Cirella GT, Zerbe S (eds) Sustainable water management and wetland restoration in settlements of continental-arid Central Asia. Bozen University Press, pp 137–155. https://doi.org/10.13124/9788860461094_10
Cirella GT, Zerbe S (2014b) Quizzical societies: a closer look at sustainability and principles of unlocking its measurability. Int J Sci Soc 5:29–45. https://doi.org/10.18848/1836-6236/CGP/v05i03/51427
Cirella GT, Tao L, Mohamed S (2007) An application of an adaptive quantitative method to measure the sustainability of the Gold Coast, Australia. J Coast Res SI50:52–56
Cirella GT, Wanjiku S, Paczoski A, Tiruneh S (2020) Human-nature relations: the unwanted filibuster. In: Cirella GT (ed) Sustainable human–nature relations: environmental scholarship, economic evaluation, urban strategies. Springer, pp 3–22. https://doi.org/10.1007/978-981-15-3049-4_1
Cirella GT, Russo A, Benassi F, Czermanski E, Goncharuk A, Oniszczuk-Jastrząbek A (2021) Energy re-shift for an urbanizing world. Energies 14:5516. https://doi.org/10.3390/en14175516
Cole MA, Rayner AJ, Bates JM (1997) The environmental Kuznets curve: an empirical analysis. Environ Dev Econ. https://doi.org/10.1017/S1355770X97000211
Czermański E, Pawłowska B, Oniszczuk-Jastrząbek A, Cirella GT (2020) Decarbonization of maritime transport: analysis of external costs. Front Energy Res 8. https://www.frontiersin.org/articles/10.3389/fenrg.2020.00028
Czermanski E, Cirella GT, Oniszczuk-Jastrząbek A, Pawlowska B, Notteboom T (2021) An energy consumption approach to estimate air emission reductions in container shipping. Energies 14. https://doi.org/10.3390/en14020278
Daly HE (2007) Ecological economics and sustainable development, selected essays of Herman Daly. https://doi.org/10.4337/9781847206947
De Bruyn SM, Van Den Bergh JCJM, Opschoor JB (1998) Economic growth and emissions: reconsidering the empirical basis of environmental Kuznets curves. Ecol Econ. https://doi.org/10.1016/S0921-8009(97)00178-X
Dobbs R, Smit S, Remes J, Manyika J, Roxburgh C, Restrepo A (2011) Urban world: mapping the economic power of cities. https://www.mckinsey.com/~/media/mckinsey/featuredinsights/urbanization/urbanworld/mgi_urban_world_mapping_economic_power_of_cities_full_report.ashx
Dogan E, Turkekul B (2016) CO2 emissions, real output, energy consumption, trade, urbanization and financial development: testing the EKC hypothesis for the USA. Environ Sci Pollut Res. https://doi.org/10.1007/s11356-015-5323-8
Fang Z, Chang Y (2016) Energy, human capital and economic growth in Asia Pacific countries - evidence from a panel cointegration and causality analysis. Energy Econ. https://doi.org/10.1016/j.eneco.2016.03.020
Fragkos P, Paroussos L (2018) Employment creation in EU related to renewables expansion. Appl Energy 230:935–945. https://doi.org/10.1016/J.APENERGY.2018.09.032
Fu Y, Yang X, Wang T, Supriyadi A, Cirella GT (2022) Spatial pattern characteristics of the financial service industry: evidence from Nanjing, China. Appl Spat Anal Policy 15. https://doi.org/10.1007/s12061-021-09409-z
Galeotti M, Lanza A (2005) Desperately seeking environmental Kuznets. Environ Model Softw. https://doi.org/10.1016/j.envsoft.2004.09.018

Gawande K, Bohara AK, Berrens RP, Wang P (2000) Internal migration and the environmental Kuznets curve for US hazardous waste sites. Ecol Econ. https://doi.org/10.1016/S0921-8009(99)00132-9
Ghosh S, Kanjilal K (2014) Long-term equilibrium relationship between urbanization, energy consumption and economic activity: empirical evidence from India. Energy 66:324–331. https://doi.org/10.1016/J.ENERGY.2013.12.052
Gonzalo J, Pitarakis JY (2002) Estimation and model selection based inference in single and multiple threshold models. J Econ 110(2):319–352. https://doi.org/10.1016/S0304-4076(02)00098-2
Gozgor G, Lau CKM, Lu Z (2018) Energy consumption and economic growth: new evidence from the OECD countries. Energy 153:27–34
Granger CWJ (1969) Investigating causal relations by econometric models and cross-spectral methods. Econometrica 37:424–438
Grossman G, Krueger A (1991) Environmental impacts of a North American free trade agreement. Natl Bureau Econ Res. https://doi.org/10.3386/w3914
Grossman GM, Krueger AB (1995) Economic growth and the environment. Q J Econ. https://doi.org/10.2307/2118443
UN Habitat (2020) World cities report 2020. The value of sustainable urbanization. Key findings and messages. https://unhabitat.org/sites/default/files/2020/11/world_cities_report_2020_abridged_version.pdf
Hadri K (2000) Testing for stationarity in heterogeneous panel data. Econ J 3:148–161
Han J, Du T, Zhang C, Qian X (2018) Correlation analysis of CO2 emissions, material stocks and economic growth nexus: evidence from Chinese provinces. J Clean Prod 180:395–406. https://doi.org/10.1016/J.JCLEPRO.2018.01.168
Hanif I (2018) Impact of economic growth, nonrenewable and renewable energy consumption, and urbanization on carbon emissions in Sub-Saharan Africa. Environ Sci Pollut Res Int 25(15):15057–15067. https://doi.org/10.1007/S11356-018-1753-4
Hansen BE (1997) Approximate asymptotic P values for structural-change tests. J Bus & Econ Stat 1, 15(1):60–67
Hansen BE (1999) Threshold effects in non-dynamic panels: estimation, testing, and inference. J Econ 93:345–368
Hansen BE (2011) Threshold autoregression in economics. Stat Interface 4:123–127
Harris RDF, Tzavalis E (1999) Inference for unit roots in dynamic panels where the time dimension is fixed. J Econ 91:201–226
He J, Richard P (2010) Environmental Kuznets curve for CO2 in Canada. Ecol Econ. https://doi.org/10.1016/j.ecolecon.2009.11.030
Hilton FGH, Levinson A (1998) Factoring the environmental Kuznets curve: evidence from automotive lead emissions. J Environ Econ Manag. https://doi.org/10.1006/jeem.1998.1023
Huang WM, Lee GWM, Wu CC (2008) GHG emissions, GDP growth and the Kyoto protocol: a revisit of environmental Kuznets curve hypothesis. Energy Policy. https://doi.org/10.1016/j.enpol.2007.08.035
Huttmanová E, Novotný R, Valentiny T (2019) An analytical view to environmental quality of life in the European Union countries. Eur J Sustain Dev 8(5):409–421. https://doi.org/10.14207/ejsd.2019.v8n5p409
Im KS, Pesaran MH, Shin Y (2003) Testing for unit roots in heterogeneous panels. J Econ. https://doi.org/10.1016/S0304-4076(03)00092-7
IPCC (2022) Climate change 2022: impacts, adaptation and vulnerability. https://www.ipcc.ch/report/ar6/wg2/
Jiao G, Lu L, Chen G, Huang Z, Cirella GT, Yang X (2021) Spatiotemporal characteristics and influencing factors of tourism revenue in the Yangtze River delta urban agglomeration region during 2001–2019. Sustainability 13:3658. https://doi.org/10.3390/su13073658
Kahraman C, Orobello C, Cirella GT (2022) Changing dynamics with COVID-19: future outlook. In: Cirella GT (ed) Human settlements: urbanization, smart sector development, and future outlook. Springer, pp 235–252. https://doi.org/10.1007/978-981-16-4031-5_13

Kaika D, Zervas E (2013) The Environmental Kuznets Curve (EKC) theory-part a: concept, causes and the CO2 emissions case. Energy Policy. https://doi.org/10.1016/j.enpol.2013.07.131

Kais S, Ben Mbarek M (2017) Dynamic relationship between CO2emissions, energy consumption and economic growth in three North African countries. Int J Sustain Energy. https://doi.org/10.1080/14786451.2015.1102910

Kao C, Chiang MH (2000) On the estimation and inference of a cointegrated regression in panel data. In: Nonstationary panels, panel cointegration, and dynamic panels. (Advances I), pp 179–222

Katircioğlu S, Katircioğlu S (2018) Testing the role of urban development in the conventional environmental Kuznets curve: evidence from Turkey. Appl Econ Lett. https://doi.org/10.1080/13504851.2017.1361004

Kourtzidis SA, Tzeremes P, Tzeremes NG (2018) Re-evaluating the energy consumption-economic growth nexus for the United States: an asymmetric threshold cointegration analysis. Energy. https://doi.org/10.1016/j.energy.2018.01.172

Kuznets S (1955) Economic growth and income inequality. Am Econ Rev 45:1–28

Kwakwa PA (2012) Disaggregated energy consumption and economic growth in Ghana. Int J Energy Econ Policy 2(1):34–40

Kwiatkowski D, Phillips PCB, Schmidt P, Shin Y (1992) Testing the null hypothesis of stationarity against the alternative of a unit root: how sure are we that economic time series have a unit root? J Econ 54:159–178

Levin A, Lin CF, Chu CSJ (2002) Unit root tests in panel data: asymptotic and finite-sample properties. J Econ. https://doi.org/10.1016/S0304-4076(01)00098-7

Lewandowski R, Cirella GT (2022) Performance management systems: trade-off between implementation and strategy development. Oper Manag Res 16. https://doi.org/10.1007/s12063-022-00305-4

Liang W, Yang M (2019) Urbanization, economic growth and environmental pollution: evidence from China. Sustain Comput: Inf Syst 21:1–9. https://doi.org/10.1016/J.SUSCOM.2018.11.007

Litavcová E, Chovancová J (2021) Economic development, CO2 emissions and energy use nexus-evidence from the Danube Region countries. https://doi.org/10.3390/en14113165

Lu Y, Yuan J, Lu X, Su C, Zhang Y, Wang C, Sweijd N (2018) Major threats of pollution and climate change to global coastal ecosystems and enhanced management for sustainability. Environ Pollut 239:670–680

Lütkepohl H (2005) New introduction to multiple time series analysis. Springer

Malhi Y, Franklin J, Seddon N, Solan M, Turner MG, Field CB, Knowlton N (2020) Climate change and ecosystems: threats, opportunities and solutions. Philos Trans R Soc B 375(1794):20190104

Mark NC, Sul D (2003) Cointegration vector estimation by panel DOLS and long-run money demand*. Oxford Bull Econ Stat 65(5):655–680. https://doi.org/10.1111/J.1468-0084.2003.00066.X

Mathews J, Tan H (2014) Economics: manufacture renewables to build energy security. Nature 513:166–168. https://doi.org/10.1038/513166a

Menegaki AN (2014) On energy consumption and GDP studies; a meta-analysis of the last two decades. Renew Sustain Energy Rev 29:31–36. https://doi.org/10.1016/J.RSER.2013.08.081

Menegaki AN, Tugcu CT (2017) Energy consumption and sustainable economic welfare in G7 countries; a comparison with the conventional nexus. Renew Sustain Energy Rev. https://doi.org/10.1016/j.rser.2016.11.133

Mensah IA, Sun M, Gao C, Omari-Sasu AY, Zhu D, Ampimah BC, Quarcoo A (2019) Analysis on the nexus of economic growth, fossil fuel energy consumption, CO2 emissions and oil price in Africa based on a PMG panel ARDL approach. J Clean Prod 228:161–174. https://doi.org/10.1016/J.JCLEPRO.2019.04.281

Miglietta PP, De Leo F, Toma P (2017) Environmental Kuznets curve and the water footprint: an empirical analysis. Water Environ J. https://doi.org/10.1111/wej.12211

Nathaniel SP, Iheonu CO (2019) CO2 abatement in Africa: the role of renewable and non-renewable energy consumption. Sci Total Environ 679:337–345. https://doi.org/10.1016/J.SCITOTENV.2019.05.011

Nathaniel SP, Alam MS, Murshed M, Mahmood H, Ahmad P (2021) The roles of nuclear energy, renewable energy, and economic growth in the abatement of CO2 emissions in the G7 countries. Environ Sci Pollut Res 28(35):47957–47972. https://doi.org/10.1007/S11356-021-13728-6
Nguyen HTT, Van Nguyen C, Van Nguyen C (2020). The effect of economic growth and urbanization on poverty reduction in Vietnam. J Asian Financ Econ Bus 7(7):229–239. https://doi.org/10.13106/JAFEB.2020.VOL7.NO7.229
Nguyen HM, Nguyen LD (2018) The relationship between urbanization and economic growth an empirical study on ASEAN countries. Int J Soc Econ 45(2):316–339. https://doi.org/10.1108/IJSE-12-2016-0358/FULL/XML
Oh W, Lee K (2004) Energy consumption and economic growth in Korea: testing the causality relation. J Policy Model 26(8–9):973–981
Olale E, Ochuodho TO, Lantz V, El Armali J (2018) The environmental Kuznets curve model for greenhouse gas emissions in Canada. J Clean Prod. https://doi.org/10.1016/j.jclepro.2018.02.178
Omri A (2014) An international literature survey on energy-economic growth nexus: evidence from country-specific studies. Renew Sustain Energy Rev 38:951–959
Oniszczuk-Jastrząbek A, Czermanski E, Cirella GT (2020) Sustainable supply chain of enterprises: value analysis. Sustainability 12:su12010419. https://doi.org/10.3390/su12010419
Ozatac N, Gokmenoglu KK, Taspinar N (2017) Testing the EKC hypothesis by considering trade openness, urbanization, and financial development: the case of Turkey. Environ Sci Pollut Res. https://doi.org/10.1007/s11356-017-9317-6
Ozcan B, Ari A (2017) Nuclear energy-economic growth nexus in OECD countries: a panel data analysis. J Econ & Manag Perspect 11(1):138–154
Özokcu S, Özdemir Ö (2017) Economic growth, energy, and environmental Kuznets curve. Renew Sustain Energy Rev. https://doi.org/10.1016/j.rser.2017.01.059
Ozturk I, Aslan A, Kalyoncu H (2010) Energy consumption and economic growth relationship: evidence from panel data for low and middle income countries. Energy Policy 38(8):4422–4428
Panayotou T (1994) Economic instruments for environmental management and sustainable development. United Nations Environment Programme's Consultative Expert Group Meeting.
Panayotou T (2003) Economic growth and the environment \r\n. Economic Survey of Europe.
Park S, Lee Y (2011) Regional model of EKC for air pollution: evidence from the Republic of Korea. Energy Policy. https://doi.org/10.1016/j.enpol.2011.06.028
Patel SH, Pinckney TC, Jaeger WK (1995) Smallholder wood production and population pressure in East Africa: evidence of an environmental Kuznets curve? Land Econ. https://doi.org/10.2307/3146715
Pearce JM (2012) Limitations of nuclear power as a sustainable energy source. Sustainability 4(6):1173–1187. https://doi.org/10.3390/SU4061173
Pedroni P (1998) Testing for convergence to common steady states in nonstationary heterogeneous panels. Indiana University
Pedroni P (2000) Fully modified OLS for heterogeneous cointegrated panels. In: Advances in econometrics. Emerald (MCB UP), Bingley, pp 93–130
Pedroni P (2001) Fully modified OLS for heterogeneous cointegrated panels. In: Nonstationary panels, panel cointegration, and dynamic panels. Emerald Group Publishing Limited, pp 93–130
Pesaran MH, Shin Y, Smith RP (1999) Pooled mean group estimation of dynamic heterogeneous panels. J Am Stat Assoc 94(446):621–634
Pesaran MH, Shin Y, Smith RJ, Hin Y, Smith RJ (2001) Bounds testing approaches to the analysis of level relationships. J Appl Econ 16(3):289–326. https://doi.org/10.1002/jae.616
Poinssot C, Bourg S, Ouvrier N, Combernoux N, Rostaing C, Vargas-Gonzalez M, Bruno J (2014) Assessment of the environmental footprint of nuclear energy systems. Comparison between closed and open fuel cycles. Energy 69:199–211. https://doi.org/10.1016/J.ENERGY.2014.02.069

Proença S, Fortes P (2020) The social face of renewables: econometric analysis of the relationship between renewables and employment. Energy Rep 6:581–586. https://doi.org/10.1016/J.EGYR.2019.09.029

Qian Y, Chakraborty TC, Li J, Li D, He C, Sarangi C, Leung L R. (2022) Urbanization impact on regional climate and extreme weather: current understanding, uncertainties, and future research directions. Adv Atmos Sci 1–42

Radmehr R, Henneberry SR, Shayanmehr S (2021) Renewable energy consumption, CO2 emissions, and economic growth nexus: a simultaneity spatial modeling analysis of EU countries. Struct Chang Econ Dyn 57:13–27. https://doi.org/10.1016/J.STRUECO.2021.01.006

Russo A, Cirella GT (2017) Biophilic cities: planning for sustainable and smart urban environments. In: Aijaz R (ed) Smart cities movement in BRICS. Observer Research Foundation and Global Policy Journal, pp 153–159

Russo A, Cirella GT (2020) Urban sustainability: integrating ecology in city design and planning. In: Cirella GT (ed) Sustainable human–nature relations: environmental scholarship, economic evaluation, urban strategies. Springer, pp 187–204. https://doi.org/10.1007/978-981-15-3049-4_10

Sachs JD, Woo WT, Yoshino N, Taghizadeh-Hesary F (2019) Importance of green finance for achieving sustainable development goals and energy security. Energy Secur 1–10

Saidi K, Hammami S (2015) The impact of CO2 emissions and economic growth on energy consumption in 58 countries. Energy Rep. https://doi.org/10.1016/j.egyr.2015.01.003

Salari M, Javid RJ, Noghanibehambari H (2021) The nexus between CO2 emissions, energy consumption, and economic growth in the U.S. Econ Anal Policy 69:182–194. https://doi.org/10.1016/J.EAP.2020.12.007

Sarkodie SA, Strezov V (2018) Empirical study of the environmental Kuznets curve and environmental sustainability curve hypothesis for Australia, China, Ghana and USA. J Clean Prod. https://doi.org/10.1016/j.jclepro.2018.08.039

Selden TM, Song D (1994) Environmental quality and development: Is there a Kuznets curve for air pollution emissions? J Environ Econ Manag. https://doi.org/10.1006/jeem.1994.1031

Shafiei S, Salim RA (2014) Non-renewable and renewable energy consumption and CO2 emissions in OECD countries: a comparative analysis. Energy Policy. https://doi.org/10.1016/j.enpol.2013.10.064

Shafik N (1994) Economic development and environmental quality: an econometric analysis. Oxf Econ Pap. https://doi.org/10.1093/oep/46.Supplement_1.757

Shahbaz M, Loganathan N, Muzaffar AT, Ahmed K, Ali Jabran M (2016) How urbanization affects CO2 emissions in Malaysia? The application of STIRPAT model. Renew Sustain Energy Rev 57:83–93. https://doi.org/10.1016/J.RSER.2015.12.096

Shiu A, Lam PL (2004) Electricity consumption and economic growth in China. Energy Policy 32(1):47–54

Spring C, Cirella GT (2022) Fostering sustainable development: green energy policy in the European Union and the United States. In: Human settlements: urbanization, smart sector development, and future outlook. Springer, pp 101–137. https://doi.org/10.1007/978-981-16-4031-5_7

Stern DI (2004) The rise and fall of the environmental Kuznets curve. World Dev. https://doi.org/10.1016/j.worlddev.2004.03.004

Stern DI, Common MS, Barbier EB (1996) Economic growth and environmental degradation: the environmental Kuznets curve and sustainable development. World Dev. https://doi.org/10.1016/0305-750X(96)00032-0

The World Bank (2020) Urban development - overview. https://www.worldbank.org/en/topic/urbandevelopment/overview#1

Timmons Roberts J, Grimes PE (1997) Carbon intensity and economic development 1962–91: a brief exploration of the environmental Kuznets curve. World Dev. https://doi.org/10.1016/s0305-750x(96)00104-0

Tong H (2012) Threshold models in non-linear time series analysis, vol 21. Springer Science & Business Media

Tsay R (2010) Analysis of financial time series. John Wiley & Sons
United Nations (2015) Transforming our world: the 2030 agenda for sustainable development. United Nations sustainable knowledge platform. Sustain Dev Goals. https://sustainabledevelopment.un.org/post2015/transformingourworld
United Nations (2019) World urbanization prospects 2018
Voelkel J, Hellman D, Sakuma R, Shandas V (2018) Assessing vulnerability to urban heat: A study of disproportionate heat exposure and access to refuge by socio-demographic status in Portland, Oregon. Int J Environ Res Public Health 15(4):640
Vural G (2020) Renewable and non-renewable energy-growth nexus: a panel data application for the selected Sub-Saharan African countries. Resour Policy 65:101568
Wang P, Bohara AK, Berrens RP, Gawande K (1998) A risk-based environmental Kuznets curve for US hazardous waste sites. Appl Econ Lett. https://doi.org/10.1080/135048598353970
Wang S, Li G, Fang C (2018) Urbanization, economic growth, energy consumption, and CO2 emissions: empirical evidence from countries with different income levels. Renew Sustain Energy Rev 81:2144–2159. https://doi.org/10.1016/J.RSER.2017.06.025
Wang Z, Ben Jebli M, Madaleno M, Doğan B, Shahzad U (2021) Does export product quality and renewable energy induce CO2 emissions: evidence from leading complex and renewable energy economies. Renew Energy 171:360–370. https://doi.org/10.1016/J.RENENE.2021.02.066
Waqih MAU, Bhutto NA, Ghumro NH, Kumar S, Salam MA (2019) Rising environmental degradation and impact of foreign direct investment: an empirical evidence from SAARC region. J Environ Manag. https://doi.org/10.1016/j.jenvman.2019.05.001
Xu SC, Miao YM, Gao C, Long RY, Chen H, Zhao B, Wang SX (2019) Regional differences in impacts of economic growth and urbanization on air pollutants in China based on provincial panel estimation. J Clean Prod 208:340–352. https://doi.org/10.1016/J.JCLEPRO.2018.10.114
Yandle B, Bhattarai M, Vijayaraghavan M (2004) Environmental Kuznets curves: a review of findings, methods, and policy implications. Research Study
Yang G, Sun T, Wang J, Li X (2015) Modeling the nexus between CO2 emissions and economic growth. Energy Policy. https://doi.org/10.1016/j.enpol.2015.06.031
Yanlong D (2018) Fiscal expenditure structure and environmental pollution: empirical research based on the panel threshold model. Meteorol Environ Res 9(1):32–37. https://doi.org/10.19547/j.issn2152-3940.2018.01.007
Yano KH, Mao KS, Wharry JP, Porterfield DM (2018) Investing in a permanent and sustainable nuclear waste disposal solution. Prog Nucl Energy 108:474–479. https://doi.org/10.1016/J.PNUCENE.2018.07.003
Zaim O, Taskin F (2000) A Kuznets curve in environmental efficiency: an application on OECD countries. Environ Resour Econ. https://doi.org/10.1023/A:1008318605976
Zhang H (2016) Exploring the impact of environmental regulation on economic growth, energy use, and CO2 emissions nexus in China. Nat Hazards 84(1):213–231. https://doi.org/10.1007/S11069-016-2417-7/TABLES/6
Zhang X, Zhang H, Yuan J (2019) Economic growth, energy consumption, and carbon emission nexus: fresh evidence from developing countries. Environ Sci Pollut Res 26(25):26367–26380
Zhao J, Dong K, Dong X, Shahbaz M (2022) How renewable energy alleviate energy poverty? A global analysis. Renew Energy 186:299–311. https://doi.org/10.1016/J.RENENE.2022.01.005
Zheng W, Walsh PP (2019) Economic growth, urbanization and energy consumption—a provincial level analysis of China. Energy Econ 80:153–162. https://doi.org/10.1016/J.ENECO.2019.01.004
Zoundi Z (2017) CO2 emissions, renewable energy and the Environmental Kuznets curve, a panel cointegration approach. Renew Sustain Energy Rev. https://doi.org/10.1016/j.rser.2016.10.018

E-Scooter Sharing and COVID-19 Impacts on the Bike Sharing Market in Polish Cities

Tomasz Bieliński and Agnieszka Ważna

Abstract Bike sharing and scooter sharing services are seen as elements that can effectively contribute to achieving the United Nations Sustainable Development Goals, particularly in terms of creating sustainable and livable cities. However, in recent years, bike sharing systems in Poland have faced a period of crisis. Introduction of e-scooter sharing services and COVID-19 created economic shocks on the micro-mobility market in cities in Poland. Bike sharing systems were shut down during the 2020 lockdown, but e-scooter sharing systems were not. This led to the fast development of e-scooter sharing which was perceived by users as safer than public transportation in terms of the threat of contagion. Most of the urban bike sharing systems recovered after these economic shocks, but e-scooter systems dominated micro-mobility market in Poland. By 2022, the e-scooter fleet was almost four times the size of bike sharing systems. In the two years from 2020 to 2022 the number of e-scooter sharing systems has quadrupled reaching 181, and the fleet size has increased fivefold reaching 94 thousand e-scooters. To compete with growing e-scooter sharing, operators of BSS in cooperation with municipal authorities should modernize their fleet to make it more attractive for users. Some of the bike sharing system operators in Polish cities have already introduced changes in their systems and improved their services by introducing new vehicles, technologies, and changes in fleet management to meet the needs of their customers.

Keywords e-scooter sharing systems · Economic shocks · New technology · Customer usage · Polish market · COVID-19 era · SDG11

T. Bieliński (✉) · A. Ważna
Faculty of Economics, University of Gdansk, 81-824 Sopot, Poland
e-mail: tomasz.bielinski@ug.edu.pl

A. Ważna
e-mail: agnieszka.wazna@ug.edu.pl

G. T. Cirella and B. Dahiya (eds.), *City Responses to Disruptions in 2020*, Advances in 21st Century Human Settlements, https://doi.org/10.1007/978-981-99-7988-2_9

1 Introduction and Literature Review

Bike sharing and scooter sharing contribute to the creation of more sustainable and livable cities and may be crucial to achieve the United Nations Sustainable Development Goal 11 (Sustainable Cities and Communities), which aims to "make cities and human settlements inclusive, safe, resilient and sustainable" (United Nations 2023). Bike sharing and scooter sharing reduce traffic congestion, noise pollution, and improve air quality. By integrating bike sharing and sustainable transportation options into urban transportation systems, cities can enhance accessibility, reduce transportation inequalities, and create more inclusive communities while also promoting the sustainability (Cirella et al. 2019; Russo and Cirella 2020; Olabi et al. 2023). This approach not only improves mobility but also aligns with the overarching goal of fostering environmentally conscious urban development, where sustainable modes of transport play a pivotal role in reducing carbon emissions and mitigating the impacts of climate change (Cirella 2020, 2022; Oniszczuk-Jastrząbek et al. 2020; Spring and Cirella 2022).

Most of the bike sharing systems (BSS) in Poland are run by private operators but depend on municipal authorities that are subsidizing their operations. COVID-19 pandemic brought decline in BSS ridership all over the world, and a crisis in the Polish BSS market. However, COVID-19 was not the only factor that led to BSS problems. We argue that there was a second reason for that crisis—the introduction and fast development of privately financed e-scooter sharing (ESS) services. Additionally, we claim that shutting down BSS during the first COVID-19 lockdown from March 31st to May 6th, 2020, led to faster migration of users to ESS and undermined efforts by the municipal authorities in providing BSS to their citizens. First lockdown in Poland due to the pandemic begun on March 16th, 2020, restrictions were eased in stages on April 24th, May 5th, May 18th, and May 30th, 2020.

The problem of competitiveness in bike sharing and e-scooter sharing was investigated only by a few studies. Some articles suggest that bike sharing, and e-scooter sharing may be used for different reasons. It was found that in Singapore e-scooter sharing is more often used for recreational reasons and tourism, while bike sharing serves for commuting (Zhu et al. 2020). This findings are consistent with that of Bieliński and Ważna (2020) based on the survey data from Tricity, Poland and McKenzie (2019) from Washington D.C., USA However, Younes et al. (2020) has found that bike sharing and e-scooter sharing may have a possible competitive relationship when casual users of bike sharing systems are considered. At the same time complementary relationship between the use of e-scooter sharing and usage of bike sharing by members (regular users) were discovered. Lee et al. (2021) forecasted that e-scooter sharing could replace 13% of bike trips (private and rented bike trips, not only bike sharing trips) in several US cities. The negative effect of the introduction of e-scooter sharing services for bike sharing was observed in Chicago where "the weekly usage of bike sharing in e-scooter sharing operation area declined by 10.2%" (Yang et al. 2021).

The study in Warsaw, Poland, revealed that there is little difference between speed of shared e-scooters and bicycles, and that placement of bicycle docks influences number of e-scooter trips which indicates that both services compete with each other. However, the study also indicated that BSS may be complementary to e-scooter sharing thanks to longer trip distances and more spatial dispersion (Nawaro 2021). Another study based on the Washington D.C. case suggests that e-scooters have both competing and complementary effects on public transport and BSS. E-scooters were also proved to be more time-efficient than public transit. Moreover, during COVID-19 pandemic, travelers were willing to pay a price premium to use e-scooters rather than using public transit (Yan et al. 2021).

Effects of COVID-19 pandemic on shared micro-mobility services was already investigated by several studies. However, as pandemic circumstances and governmental polices changed over time, they created new challenges for scholars. It has been well established that lockdown periods decreased overall mobility which also affected bike sharing and e-scooter sharing (Li et al. 2021). Strict lockdown policies affected micro-mobility in Chinese cities like Beijing (Shang et al. 2021), where usage of bike sharing decreased by 60% compared to the same period in 2019 (Chai et al. 2020). In Nanjing, China, station-based bike sharing trips were reduced by 72% and free-floating bike sharing trips by 82% (Hua et al. 2021).

Lockdowns also affected distribution of trip purposes. While the share of shopping and leisure trip destinations decreased, percent of trips that end in groceries, parks and homes increased (Li et al. 2021). COVID-19 pandemic also created a shift in the motivations for using BSS. In Lisbon, Portugal, people started using shared bicycles to avoid public transport and to maintain a social distance (Teixeira et al. 2021). Wang and Noland (2021) found that in the New York City subway ridership dropped by 95%, and bike sharing suffered 70% trip reduction during the pandemic peak period in 2020. BSS ridership in New York City gradually recovered to the pre-pandemic levels, but subway trips were 30% below the levels registered before the outbreak of COVID-19. Moreover, the study found the number of longer trips (between 30 and 60 min) in BSS increased, which suggested that some of them shifted from the subway. This kind of COVID-19 shock results were not possible in Poland, where BSS were closed by the decision of the Polish government.

2 Bike Sharing and E-Scooter Sharing Systems in Poland Before the COVID-19 Pandemic

The first BSS in Poland was introduced in Cracow in 2008. The most dynamic growth of BSS' popularity in the country took place one decade later. The number of BSS increased from 31 in 2017 to 56 in 2018 (by over 80%) and so increased the number of active users (by 39.1%), trips (by 28.7%) and the fleet size (by 14.5%) (Bieliński et al. 2019). Number of BSS in Poland reached their peak in 2019, when there were 96 available systems in almost all large- and medium-sized cities. Table 1 presents the descriptive statistics concerning BSS in Poland three years before the COVID-19 pandemic.

The first e-scooter sharing system in Poland was introduced in October 2018 in Wroclaw (Elektrowoz 2022). At that time, Poland was the sixth country in Europe to start off the e-scooter sharing. E-scooter sharing systems, being privately financed and operating independently of local government authorities, spread all over the country in 2019. In total, there were 19 cites (Bialystok, Bydgoszcz, Czestochowa, Gdansk, Gliwice, Kalisz, Katowice, Kolobrzeg, Krakow, Lublin, Lodz, Opole, Plock, Poznan, Radom, Szczecin, Warsaw, Wroclaw and Zielona Gora) with e-scooter sharing systems including 18 cities (except Gdansk) where bike and e-scooter sharing systems coexisted (Mobilne Miasto 2020). The estimated number of e-scooters available in Polish cities in 2019 was 10,000 (Mobilne Miasto 2021). Besides the biggest operator, the Lime company, entities providing shared e-scooters in Poland in 2019 were: Bird, Hive, blinkee.city and citybee (Miastojestnasze.org 2019).

Introduction of e-scooter sharing systems in the biggest Polish cities coincided with a first breakdown on the bike sharing market after the years of undisturbed expansion and growing popularity since 2008. Two big BSSs in the city of Cracow and Tricity (cities of Gdansk, Gdynia, Sopot and neighboring municipalities) were shut down suddenly in 2019 due to unprofitability. Although the total number of working BSS (56) in Poland in 2019 was at its peak, there was a significant decline in number of bicycle rentals within 15 biggest BSS (by 19%). The same year, the number of bicycles available within these 15 systems rose on average by 5% (Mobilne Miasto 2020), which was expected to bring additional users and rides. As other conditions

Table 1 Bike sharing systems in Poland in 2017–2019

Year	Number of BSS	Number of trips/rentals	Number of active/registered users	The fleet size (number of bicycles)	Most significant changes on the market
2017	31	12 million	1.3 million	16 000	–
2018	56	15.5 million	1.8 million	18 000	- October: first e-scooter sharing system opened
2019	96	24% lower number of average monthly rentals per bike (in 15 biggest systems)*	2.6 million	26 632	- Two big lockdowns - Decline in number of rentals in 15 biggest systems - Expansion of e-scooter sharing

* Lack of annual data for all systems in 2019

Source Authors' own elaboration based on Bieliński et al. (2019) and Mobilne Miasto (2020)

(infrastructure, legal environment, etc.) remained unchanged, the decline in ridership could be associated with the appearance of shared e-scooters.

3 Bike Sharing and e-scooter Sharing Systems in Poland During the COVID-19 Pandemic

During 2020–2021, the number of Polish cities with functioning ESS rose rapidly despite the COVID-19 pandemic, i.e., from almost 20,000 vehicles in 2020 to 46,182 vehicles in 2021 (Smartride.pl 2022). At that time, there were no restrictions concerning the e-scooters usage due to the impact of COVID-19. The situation on the bike sharing market was quite the opposite as the first lockdown in Poland resulted in the suspension of BSS from 31st March to 6th May 2020 (the very beginning of Polish spring and cycling season). This decision was applicable nationwide on the basis of a regulation of the Council of Ministers and what is important is that it did not refer to other vehicles for rent, such as e-scooters, mopeds, or cars (gs24.pl 2020). In 2020, BSS in Polish cities declined in terms of the number of available vehicles by 33% (Mobilne Miasto 2021). It was the biggest decrease among all shared micro-mobility services that year (Fig. 1).

Temporary shutdown of BSS was also in force in Spanish cities such as Madrid and Saragossa. However, in many other metropolises around the world (such as Vienna, Budapest, London, Paris, many German cities, Bogota, New York, Chicago, and Philadelphia), local governments decided to promote bike sharing systems as a safe means of transportation during the COVID-19 pandemic (Turoń and Kubik 2021).

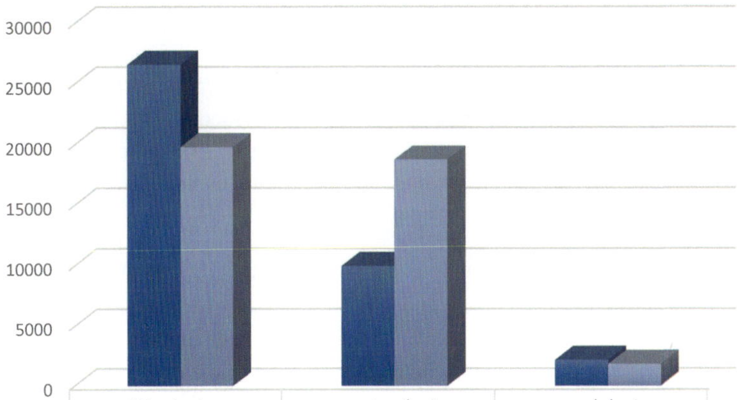

	bike sharing	e-scooter sharing	moped sharing
2019	26632	10000	2180
2020	19834	18771	1836

Fig. 1 Polish micro-mobility fleet size in 2019–2020. *Source* Authors' own elaboration based on Smartride.pl (2022)

Governments supported BSS operators in encouraging passengers to use bike sharing services, for example, by introducing fee waivers, and enabling and facilitating the usage of urban infrastructure to ride a bike. These solutions worked, for example, the Chicago BSS recorded two-fold increase in the number of users in March 2020 compared to the previous year (philadelphia.cbslocal.com 2020).

In years of pandemic (2020 and 2021), privately financed ESS in Poland developed fast, while BSS struggled, but eventually survived. In the Q3 (peak months) of 2021, the fleet of shared bicycles accounted for 21.7 thousand in 75 systems. It was higher than in 2020, but lower than in pre-COVID-19 2019, when bicycle fleet peaked at 26.6 thousand in 96 systems. The number of BSS operators decreased, as two companies went out of the market. Solow recovery of BSS continued in 2022 when the number of running systems increased to 82 and bicycle fleet size raised to 23.7 thousand (Table 2).

Given data on Polish micro-mobility during and after the COVID-19 pandemic compared to the data concerning pre-pandemic years show that e-scooter sharing market dominates and develops the faster within shared micro-mobility services. In 2021, the e-scooter sharing fleet size was more than twice bigger than the bike sharing systems' fleet size and in 2022 it was almost 4 times as big (Table 2). Also, ESS were introduced in smaller cities and in 2022 for the first time there were more ESS (161 systems) than BSS (82 systems). The leading company on the e-scooter sharing market in 2022 was Bolt with the fleet of 34.7 thousand vehicles operating in 34 cities (37% market share). The Lime company was the second biggest operator with 21.6 thousand e-scooters in 30 cities (23% market share). Other important operators were TIER and Dott having accordingly 21% and 13% market share (Smartride.pl 2022).

Bike sharing systems in Poland in 2021 put into service 9% more vehicles compared to the previous year. After hard time of the first pandemic months in 2020, the BSS market managed to survive and, in some cities, extended the offer in 2021. At the same time operators also encountered some problems in other localizations. In 2021, Nextbike had the biggest share of the BSS market (75%) with 37 systems and 16.188 bicycles. The rest of the market (25%) was divided by the following operators: Freebike (12% with 2.680 bikes), Roovee (9% with 1.910 bikes), GeoVelo (3% with 613 bikes), Filfri Bikes (1% with 250 bikes) and Blinkee.city (0,1% with 18

Table 2 BSS and ESS in 2020–2021 in Polish Cities–data as of the Q3 (peak months) of each year. Peak months (end of Q3/September)

Year	Number of running systems		Number of operators		The fleet size	
	BSS	ESS	BSS	ESS	BSS	ESS
2020	69	39	8	12	19,834	18,771
2021	75	60	6	11	21,659	46,182
2022	82	161	7	12	23,700	94,000

Source Authors' own elaboration based on Smartride.pl (2022)

bikes) (Smartride.pl 2022). It is also worth mentioning that in 2021 Nextbike has been acquired by micro-mobility operator Tier, one of the biggest e-scooter sharing companies in the global market (Ułan 2021).

The biggest system in Poland, Veturilo in Warsaw, began the 2021 season with reductions in the number of stations and vehicles as shown in Table 3. Reductions were also made in Bialystok (10 stations and 80 bicycles less). In Wroclaw, the second biggest BSS in 2021 there were no changes in fleet size or a number of stations, the system was also the only one running all year round without a winter break. The third biggest system runs in Poznan with an extended offer (new free-floating and electric bikes). The Lublin BSS also developed (new parking space for free-floating vehicles) and so did the BSS in Katowice (more stations) and Bydgoszcz (more stations and vehicles). Problems with re-opening the BSS in 2021 occurred in Szczecin and Lodz (in this case also in 2020) as shown in Table 3.

In Cracow, the second biggest city in Poland, the oldest BSS in the whole country collapsed in 2019. In the official announcement, BSS operator explained that the reason was the unsuccessful business model and growing popularity of ESS that were introduced in the same year (Janoś 2019). Cracow's BSS business model assumed a contract with the city, but with relatively small municipal subsidies. ESS operates without any contract within the city, and thus without the requirements and standards that a bicycle operator must adhere to (Dybalski 2019). One of the largest cities in Poland, where the BSS collapsed and has not been restored was Gdansk. It is impossible to determine if the introduction of ESS had any connection with the bankruptcy of the local system operator (NB Tricity), as the system was experimental (it was the only system in Poland with the fleet of 100% electrically assisted bicycles) and functioned only for seven months in 2019 (Bieliński et al. 2021). ESS was introduced in Gdansk in the same year as BSS (Bieliński and Ważna 2020) and developed rapidly after the bankruptcy of the BSS. In Lodz, BSS runs for three years from 2016 to 2019. It was closed in 2020 officially due to the COVID-19 pandemic and has not been functioning till the end of July 2021. The new operator opened the system with a 74-day delay and provided only 300 bicycles on 30 stations instead of the planned amount of 1500 vehicles on 150 stations (Andrzejewski 2021; rowerowalodz.pl 2021). BSS in Szczecin after seven years (2014–2020) of operation was closed in 2021 as authorities decided to modernize the system and change the operator at the same time. In 2020, the number of bicycle rentals (trips) in Szczecin decreased after pandemic restrictions and e-scooter sharing introduction (215.000 rentals in 2020 comparing to the record number of 565.000 rentals in 2016) (Wasilewski 2021). As a consequence, changes in the BSS market were inevitable. The new system is planned to be free-floating and equipped with 800 new vehicles available all year round (Nawrocka 2022). The old fleet was put up for sale at low prices at the end of 2021 (Brzeziński 2021).

Table 3 Bike sharing systems in the biggest Polish cities in 2021

The biggest cities (over 250 000 citizens)	Population (in thousands)	Number of stations in 2021	Number of bicycles in 2021	Important changes comparing to the previous years	Year of ESS introduction
Warsaw	1794.2	300	4500	Reduction: • 100 less stations and 1200 less bicycles comparing to 2020 • Even 2000 less bicycles than in previous years	2018
Cracow	780.0	First BSS in Poland, operating 2008–2019 Stations: 169; Bicycles: 1500			2019
Lodz	672.2	Operated in 2016–2019, than closed due to COVID-19 pandemic. In 2021 new operator emerged and opened the system partially with 74 days of delay. The target size of planned new system: 150 stations, 1500 bicycles			2019
Wroclaw	641.9	200	2065	Development: • 100 more bicycles (electric)	2018
Poznan	532.0	113	1700	Development: • Free-floating system • Electric bicycles • User-friendly fee changes	2018
Gdansk	470.8	First fully electric BSS in Poland operating only in 2019 Stations: 660; Bicycles: 1400; System partially free-floating			2019
Szczecin	398.3	Operated in 2014–2020. Under modernization, to be re-opened in 2022 as a free-floating system with about 800 bicycles			2019

(continued)

Table 3 (continued)

The biggest cities (over 250 000 citizens)	Population (in thousands)	Number of stations in 2021	Number of bicycles in 2021	Important changes comparing to the previous years	Year of ESS introduction
Bydgoszcz	344.1	56	600	Development: • 4 more stations • 10 more bicycles	2019
Lublin	338.6	130	700	Development: • New parking space for free-floating BSS	2019
Bialystok	297.0	54	570	Reduction: • 10 less stations • 80 less bicycles	2019
Katowice	290.6	105	660	Development: • About 20 stations more	2019

Source Authors' own elaboration

4 Concluding Remarks

COVID-19 created an economic shock that affected urban mobility all over the world, and decreased number in bike sharing trips in Poland was widely seen as its consequence (Dzięcielski et al. 2020; Suchanek et al. 2021; Kahraman et al. 2022). However, our study shows that there could have been another factor that contributed to this crisis. It was an introduction of competing ESS financed by private companies. This crisis was aggravated by the COVID-19 pandemic that led to further destabilization of the market, but problems of BSS in Poland begun before the spread of the virus. Lockdown of BSS imposed by the Polish government led to fleet reductions and closure of operations by many BSS. As ESS services were not restricted, they significantly increased the scale of their operations despite considerably limited urban mobility. This way privately financed ESS was favored against publicly co-financed BSS. The BSS market managed to endure the crisis, but we do not recommend introducing this kind of selectively targeted restrictions for the future as it can cause shocks to the market and there is no evidence that this could limit the spread of viruses. However, certainly all forms of micro-mobility can allow for easier social distancing than public transport (Hosseinzadeh and Kluger 2021). Furthermore, the situation on the Polish micro-mobility market suggests that BSS can coexist with ESS and grow despite its competition. However, to compete BSS should modernize its fleet and introduce new features (like for example free-floating bicycles, cargo bicycles, electrically assisted bikes), as it was carried out in the Polish cities like Wroclaw, Poznan, Lublin, and Szczecin.

References

Andrzejewski T (2021) Startuje Łódzki Rower Publiczny 2.0 - Urząd Miasta Łodzi. https://uml.lodz.pl/aktualnosci/artykul/startuje-lodzki-rower-publiczny-20-id42481/2021/7/30/. Accessed 31 Jan 2022

Bieliński T, Ważna A (2020) Electric scooter sharing and bike sharing user behaviour and characteristics. Sustainability (Switzerland) 12. https://doi.org/10.3390/su12229640

Bieliński T, Kwapisz A, Ważna A (2019) Bike-sharing systems in Poland. Sustainability (switzerland) 11(9):2458. https://doi.org/10.3390/su11092458

Bieliński T, Kwapisz A, Ważna A (2021) Electric bike-sharing services mode substitution for driving, public transit, and cycling. Transp Res D Transp Environ 96. https://doi.org/10.1016/j.trd.2021.102883

Brzeziński M (2021) Szczecin sprzedaje rowery miejskie za półdarmo. https://www.auto-swiat.pl/wiadomosci/aktualnosci/szczecin-sprzedaje-rowery-miejskie-za-poldarmo/hrd9g8p. Accessed 31 Jan 2022

Cirella GT (2020) Sustainable human-nature relations: environmental scholarship: economic evaluation, urban strategies. Springer, Berlin. https://doi.org/10.1007/978-981-15-3049-4

Cirella GT (2022) Human settlements: urbanization, smart sector development, and future outlook. Springer. https://doi.org/10.1007/978-981-16-4031-5

Cirella G, Bak M, Koźlak A, Pawlowska B, Borkowski P (2019) Transport innovations for elderly people. Res Transp Bus Manag 30:100381. https://doi.org/10.1016/j.rtbm.2019.100381

Chai X, Guo X, Xiao J, Jiang J (2020) Analysis of spatial-temporal behavior pattern of the share bike usage during COVID-19 pandemic in Beijing

Dybalski J (2019) Kraków. BikeU wypowiedziało umowę na Wavelo. Zmierzch rowerów publicznych? Transport Publiczny. https://www.transport-publiczny.pl/wiadomosci/krakow-bikeu-porzuca-wavelo-zmierzch-rowerow-publicznych-63136.html. Accessed 31 Jan 2022

Dzięcielski M, Woźniak M, Radzimski A (2020) Exploring bike-sharing systems in Poland co-financed from the European Union funds. Prace Komisji Geografii Komunikacji PTG Nr 23(4):7–18. https://doi.org/10.4467/2543859XPKG.20.023.13125

Elektrowoz.pl Wrocław z elektrycznymi hulajnogami Lime. CENNIK: 2 zł odblokowanie + 0,5 zł/min. [aktualizacja] • SAMOCHODY ELEKTRYCZNE – www.elektrowoz.pl. https://elektrowoz.pl/ro-hu-de/wroclaw-z-elektrycznymi-hulajnogami-lime-cennik-2-zl-odblokowanie-0-5-zl-min-aktualizacja/. Accessed 19 Jan 2022

gs24.pl (2020) Rowery miejskie wyłączone z użytku, a co ze skuterami i hulajnogami? Są dezynfekowane|Głos Szczeciński. https://gs24.pl/rowery-miejskie-wylaczone-z-uzytku-a-co-ze-skuterami-i-hulajnogami-sa-dezynfekowane/ar/c1-14897727. Accessed 20 Jan 2022

Hosseinzadeh A, Kluger R (2021) Analyzing the impact of COVID-19 pandemic on micromobility transportation. In: International conference on transportation and development 2021: transportation planning and development - selected papers from the international conference on transportation and development 2021, pp 52–60. https://doi.org/10.1061/9780784483541.005

Hua M, Chen X, Cheng L, Chen J (2021) Should bike-sharing continue operating during the COVID-19 pandemic? Empirical findings from Nanjing, China. J Transp Health 23:101264. https://doi.org/10.1016/J.JTH.2021.101264

Janoś K (2019) Rower miejski w kryzysie. Hulajnogi elektryczne i wysokie koszty jak kij w szprychy - Money.pl. https://www.money.pl/gospodarka/rower-miejski-w-kryzysie-hulajnogi-elektryczne-i-wysokie-koszty-jak-kij-w-szprychy-6448805530379905a.html. Accessed 31 Jan 2022

Lee M, Chow JYJ, Yoon G, He BY (2021) Forecasting e-scooter substitution of direct and access trips by mode and distance. Transp Res D Transp Environ 96:102892. https://doi.org/10.1016/J.TRD.2021.102892

Li A, Zhao P, Haitao H et al (2021) How did micro-mobility change in response to COVID-19 pandemic? A case study based on spatial-temporal-semantic analytics. Comput Environ Urban Syst 90:101703. https://doi.org/10.1016/J.COMPENVURBSYS.2021.101703

Kahraman C, Orobello C, Cirella GT (2022) Changing dynamics with COVID-19: future outlook. In Cirella GT (ed) Human settlements: urbanization, smart sector development, and future outlook. Springer, pp 235–252. https://doi.org/10.1007/978-981-16-4031-5_13

McKenzie G (2019) Spatiotemporal comparative analysis of scooter-share and bike-share usage patterns in Washington, D.C. J Transp Geogr 78:19–28. https://doi.org/10.1016/J.JTRANGEO.2019.05.007

Miastojestnasze.org (2019) e-hulajnoga – nowy środek transportu? — Miasto Jest Nasze. https://miastojestnasze.org/e-hulajnogi-analiza-szansy-zagrozenia/. Accessed 20 Jan 2022

Mobilne Miasto (2020) Ostre hamowanie roweru miejskiego. Bike sharing w Polsce 2019/2020. Epizod, czy początek kryzysu? https://smartride.pl/wp-content/uploads/2020/03/Mobilne_Miasto_raport_bikesharing_final.pdf. Accessed 19 Jan 2022

Mobilne Miasto (2021) Sress test zaliczony. Mikromobilność w Polsce 2020. https://smartride.pl/wp-content/uploads/2021/01/Raport_MM_-Mikromobilnosc_-2020_w_Polsce_FINAL.pdf. Accessed 19 Jan 2022

Nawaro Ł (2021) E-scooters: competition with shared bicycles and relationship to public transport. https://doi.org/10.1080/19463138202111981336

Nawrocka K (2022) Nowe rowery miejskie są już w Szczecinie. Kiedy mieszkańcy będą mogli się nimi przejechać? Szczecin. https://wszczecinie.pl/aktualnosci,nowe_rowery_miejskie_sa_juz_w_szczecinie_kiedy_mieszkancy_beda_mogli_sie_nimi_przejechac,id-40009.html. Accessed 31 Jan 2022

Nowy Jork. Mieszkańcy zamienili metro na rowery w obawie przed koronawirusem - Transport Publiczny. https://www.transport-publiczny.pl/wiadomosci/nowy-jork-mieszkancy-zamienili-metro-na-rowery-w-obawie-przed-koronawirusem-64127.html. Accessed 20 Jan 2022

Olabi AG, Wilberforce T, Obaideen K et al (2023) Micromobility: progress, benefits, challenges, policy and regulations, energy sources and storage, and its role in achieving sustainable development goals. Int J Thermofluids 17:100292. https://doi.org/10.1016/J.IJFT.2023.100292

Oniszczuk-Jastrząbek, A., Czermanski, E., & Cirella, G. T. (2020). Sustainable Supply Chain of Enterprises: Value Analysis. Sustainability, 12, su12010419. https://doi.org/10.3390/su12010419

philadelphia.cbslocal.com (2020) Coronavirus Philadelphia: Indego bike share offering $5 monthly pass to help community during COVID-19 pandemic – CBS Philly. https://philadelphia.cbslocal.com/2020/04/27/coronavirus-philadelphia-indego-bike-share-offering-5-monthly-pass-help-community-during-covid-19-pandemic/. Accessed 20 Jan 2022

rowerowalodz.pl (2021) Startuje rower publiczny. https://rowerowalodz.pl/rower-publiczny/startuje-rower-publiczny. Accessed 31 Jan 2022

Russo A, Cirella GT (2020) Urban sustainability: integrating ecology in city design and planning. In: Cirella GT (ed) Sustainable human–nature relations: environmental scholarship, economic evaluation, urban strategies. Springer, pp 187–204. https://doi.org/10.1007/978-981-15-3049-4_10

Shang WL, Chen J, Bi H et al (2021) Impacts of COVID-19 pandemic on user behaviors and environmental benefits of bike sharing: a big-data analysis. Appl Energy 285:116429. https://doi.org/10.1016/J.APENERGY.2020.116429

Smartride.pl (2022) Mikromobilność - Strefa Danych. SmartRide.pl i Mobilne Miasto. https://smartride.pl/strefa-danych/. Accessed 20 Jan 2022

Spring C, Cirella GT (2022) Fostering sustainable development: green energy policy in the European Union and the United States. In: Human settlements: urbanization, smart sector development, and future outlook. Springer, pp 101–137. https://doi.org/10.1007/978-981-16-4031-5_7

Suchanek M, Jagiełło A, Suchanek J (2021) Substitutability and complementarity of municipal electric bike sharing systems against other forms of urban transport. Appl Sci 11:6702. https://doi.org/10.3390/APP11156702

Teixeira JF, Silva C, Moura e Sá F (2021) The motivations for using bike sharing during the COVID-19 pandemic: Insights from Lisbon. Transp Res Part F Traffic Psychol Behav 82:378–399. https://doi.org/10.1016/J.TRF.2021.09.016

Turoń K, Kubik A (2021) Business innovations in the new mobility market during the COVID-19 with the possibility of open business model innovation. J Open Innov: Technol Market Complex 7:195. https://doi.org/10.3390/JOITMC7030195

Ułan G (2021) Nextbike przejęte przez TIER! To teraz największy operator mikromobilności w Europie. https://antyweb.pl/tier-przejmuje-nextbike. Accessed 1 Feb 2022

United Nations (2023) Goal 11|Department of Economic and Social Affairs. https://sdgs.un.org/goals/goal11. Accessed 26 Jun 2023

Wang H, Noland RB (2021) Bikeshare and subway ridership changes during the COVID-19 pandemic in New York City. Transp Policy (oxf) 106:262–270. https://doi.org/10.1016/J.TRANPOL.2021.04.004

Wasilewski S (2021) Dlaczego stacje roweru miejskiego są puste? Kiedy ruszy Bike_S w Szczecinie? Szczecin Nasze Miasto. https://szczecin.naszemiasto.pl/dlaczego-stacje-roweru-miejskiego-sa-puste-kiedy-ruszy/ar/c1-8161759. Accessed 1 Feb 2022

Yang H, Huo J, Bao Y et al (2021) Impact of e-scooter sharing on bike sharing in Chicago. Transp Res Part A Policy Pract 154:23–36. https://doi.org/10.1016/J.TRA.2021.09.012

Yan X, Yang W, Zhang X et al (2021) A spatiotemporal analysis of e-scooters' relationships with transit and station-based bikeshare. Transp Res D Transp Environ 101:103088. https://doi.org/10.1016/J.TRD.2021.103088

Younes H, Zou Z, Wu J, Baiocchi G (2020) Comparing the temporal determinants of dockless scooter-share and station-based bike-share in Washington, D.C. Transp Res Part A Policy Pract 134:308–320. https://doi.org/10.1016/J.TRA.2020.02.021

Zhu R, Zhang X, Kondor D et al (2020) Understanding spatio-temporal heterogeneity of bike-sharing and scooter-sharing mobility. Comput Environ Urban Syst 81:101483. https://doi.org/10.1016/J.COMPENVURBSYS.2020.101483

Impact of the Global Pandemic on the Activities and Preparedness of Protection, Rescue, and Relief Services in Slovenia

Janja Kramer Stajnko, Giuseppe T. Cirella, and Matjaž P. Nekrep

Abstract Safety of the state and citizens is based on an efficient system of protection against natural and other disasters and is implemented by protection, rescue, and relief forces. Today, the volunteer firefighting organizations in Slovenia are the largest and most efficient part of the system of protection, rescue, and relief forces. The high operational capability and good equipment of the firefighting units enable them to be on standby 24 h a day to respond to natural and other disasters throughout the country. The crisis of the COVID-19 pandemic and the lockdown had a massive impact on the activities and preparedness of the protection, rescue, and relief forces, which will have long-term consequences. Due to the lockdown in 2020, all activities such as education and training were disrupted, which are essential to maintaining preparedness for emergencies. In addition, the disruption of various social activities that are essential for building and sustaining the system could lead to a lack of volunteers in the near future. It is therefore important that governments, municipalities, and other local authorities immediately direct their action plans toward measures that will restore the operational capacity of these forces.

Keywords Safety · Natural disasters · Prevention · Disaster management · Operational capability · Firefighting · Emergency services · Volunteers

J. K. Stajnko (✉) · M. P. Nekrep
Faculty of Civil Engineering, Transportation Engineering and Architecture, University of Maribor, 2000 Maribor, Slovenia
e-mail: janja.kramer@um.si

M. P. Nekrep
e-mail: matjaz.nekrep@um.si

G. T. Cirella
Faculty of Economics, University of Gdansk, 81-824 Sopot, Poland
e-mail: gt.cirella@ug.edu.pl

G. T. Cirella and B. Dahiya (eds.), *City Responses to Disruptions in 2020*, Advances in 21st Century Human Settlements, https://doi.org/10.1007/978-981-99-7988-2_10

1 Introduction

The COVID-19 pandemic has transformed many aspects of human life and the economy, leaving massive consequences throughout society. The most basic foundations of social well-being have been affected, such as income and wealth, work and job quality, housing, subjective well-being, social relationships, work-life balance, environmental quality, and security (Cirella 2020, 2022; Cirella et al. 2020, 2021; Fu et al. 2022; Kahraman et al. 2022). An efficient disaster management system is essential in any society to reduce the number of disasters and to prevent or reduce the number of victims and other consequences of such disasters. In Slovenia, a comprehensive system of protection against natural and other disasters is organized by the state, municipalities, and other local communities to implement the protection of people, animals, property, cultural heritage, and the environment against natural and other disasters. It is a unified subsystem of the national security system that is orchestrated and linked with other national security subsystems at all levels. It includes programming, planning, organizing, implementing, monitoring, and financing measures and activities to protect against natural and other disasters (ACPDR 2022).

Due to Slovenia's geographical features, natural disasters, especially floods, summer storms, fires, and earthquakes, are a frequent phenomenon. The risk of environmental disasters is increasing due to urbanization and industrialization (Cirella et al. 2020, 2021; Cirella and Zerbe 2014; Fu et al. 2022; Jiao et al. 2021; Lewandowski and Cirella 2022; Oniszczuk-Jastrząbek et al. 2020; Russo and Cirella 2020; Spring and Cirella 2022); however, the current threat of terrorism and the emergence of various contagious diseases also contribute significantly. The annual taxes paid by Slovenia due to natural and other disasters amount on average to more than 2% of GDP (ACPDR 2022).

The strongest force in the system of protection, rescue, and relief forces are the volunteer firefighting units organized throughout the country. The Slovenian firefighting organization is one of the oldest and most popular in the whole country and is recognized as one of the most developed in Europe and in the world, both in terms of its multitude operational capability and its proportion of young people and women. Of Slovenia's 2,000,000 inhabitants, more than 165,000 are members of a firefighting organization, which represents around 7.7% of the population, of which 50,000 are operative firefighters (between 18 years of age and around 63 for men and 50 for women), ready to intervene in an emergency around the clock (Kramer Stajnko 2019; Firefighting Association of Slovenia 2020). For comparison, the membership in volunteer fire departments in Austria and the Czech Republic is around 4%, Croatia 3%, and Germany 1.7% (Kladnik 2020a). Professional fire departments, however, are only organized in larger cities—across the country there are thirteen professional fire departments with about 800 professional firefighters. In case of disasters, professionals and volunteers work together, as the capacities of professional units are not sufficient to cover large disasters and accidents.

Volunteer firefighting organization in Slovenia is, besides their role as associations of public safety, important in view of their social, and potentially political roles, since throughout history, especially local politicians, often build their career through activities in volunteer associations (Kladnik 2020a, b; Ratej 2013). As such, volunteer firefighters have demonstrated a strong adherence to their voluntary tradition and independence from any government structure (Kladnik 2020a). To ensure the proper functioning of the volunteer fire department system, it is necessary to constantly work on various areas in addition to education and training, where firefighters learn the basics of rescue, equipment, first aid, etc., in order to be efficient in the event of an emergency. In recent decades, it has been found that volunteering is a major problem especially in urban areas as people are more involved in their work, have less social life, and less time to serve the social community (Leskovec and Kramer Stajnko 2021). However, the COVID-19 pandemic exacerbated the crisis of volunteering in all areas as most social activities had been curtailed or blocked during the lockdown in 2020.

This chapter analyzes how the global COVID-19 pandemic has affected the activities of volunteer firefighters during the first two years of crisis, from different perspectives: education and training of volunteer firefighters, social life in fire units, and expected long-term impact on the work of these organizations. Based on the conclusions, some necessary actions are suggested for the main leaders to maintain the full functioning of firefighting units and ensure social well-being.

2 System of Protection Against Natural and Other Disasters in Slovenia

Protection against natural and other disasters in Slovenia is the responsibility of the citizenry of the Republic of Slovenia as individuals volunteer in associations, professional bodies, and other non-governmental organizations dealing with protection against natural and other disasters, public rescue services, commercial companies, institutes and other organizations, local communities, and the state (ACPDR 2022). The objective of the system can be represented in the form of a disaster management cycle (Fig. 1).

In Slovenia, the system of protection against natural and other disasters is coordinated by the Administration of the Republic of Slovenia for Civil Protection and Disaster Relief (ACPDR) which is a constituent body of the Ministry of Defense. It performs administrative and professional protection, rescue, and relief tasks as well as other tasks related to the field, e.g., management, preparation, and operation of the system for protection against natural and other disasters, the provision of assistance to affected local communities, monitoring, notification and warning, communications and the information system, the assessment of damage caused by natural and other disasters, fire protection and the operation of general rescue services, the organization and operation of the civil protection service and the preparation and operation of the civil protection headquarters, the preparation and implementation of national

Fig. 1 Disaster management cycle and the main objectives of the system of protection against natural and other disasters (Kramer Stajnko and Petek 2019)

programs in the field of protection from natural and other disasters, and disaster relief and post-disaster remediation (ACPDR 2022).

When executing the tasks of protection, rescue, and relief in the event of natural and other disasters, the Administration cooperates the Slovenian Armed Forces, the police, and other services. It also leads, coordinates, and implements the international activities of the country in the field of rescue and protection at bilateral and multilateral levels.

3 Protection, Rescue, and Relief Forces

Protection, rescue, and relief forces are the capabilities of the state, local communities, companies, institutes, or other organizations that are available for protection, rescue, and relief in the event of natural or other disasters. They are intended for: first aid and emergency medical assistance, assistance to the affected population and population at risk, first veterinary assistance, extinguishing fires and fire rescue, rescue from debris, landslides and avalanches, rescue in the event of floods or other meteorological disasters, environmental and other accidents on sea, rivers and lakes, rescue in the event of major traffic accidents, rescue in the event of mining accidents, mountain rescue, cave rescue, water rescue, protection against unexploded ordnance, protection against avalanches, radiological, chemical and biological protection against nuclear, chemical or biological weapons, other means of mass destruction, and in the event of accidents involving dangerous substances, searching for missing persons in the event of natural and other disasters, rescue in the event of war, terrorist attacks and other forms of mass violence, and provision of basic living conditions. Protection, rescue, and relief forces must be organized by the state, local communities and certain companies, and institutes and other organizations, based on the risk in the activities they perform. In the field of disaster response, the Republic of Slovenia is divided into thirteen regions, which are covered by regional offices of ACPDR (Ratchev 2015).

Depending on the nature of citizen involvement and participation, protection, rescue, and relief forces are divided into volunteer and professional forces, and response forces. Individual units and services may also be organized to combine professional and volunteer members. Voluntary units and emergency services are organized in non-governmental organizations, especially humanitarian organizations. Their activities are complemented by professional emergency services. Professional units and services for protection, rescue, and relief are independent units or services that operate in the field of protection and rescue when needed. Mandatory units and services for protection, rescue, and assistance are organized as civil defense units and services on the basis of civic duty. Protection, rescue, and relief tasks are carried out by:

- units, services, and other operational structures of societies and organizations performing protection, rescue, and assistance tasks or public services on the basis of a decision by the competent national local authority (i.e., firefighting units and units and services of associations and non-governmental organizations);
- companies, institutes, and other organizations coordinating rescue units and services on the basis of a decision by the competent local or national authority and taking into account the risk associated with the activities they carry out;
- civil protection units and services are organized on the basis of civic duty as additional protection, rescue, and relief forces (i.e., established by the state, local communities and enterprises, institutes, and other organizations in accordance with the criteria for the organization, equipment, and training of protection, rescue, and relief forces);
- the police, which provides security, law and order and participates in rescue operations with helicopters, depending on the availability of its forces; and
- the Slovenian Armed Forces with its air unit, nuclear, biological and chemical defense unit, technical unit and medical service, and other units when not involved in defense tasks.

4 Voluntary Firefighting Organization in Slovenia

The organized form of voluntary firefighting in Slovenia has a 150-year tradition. Today, voluntary firefighting organizations in Slovenia are the largest and most efficient part of the system of protection, rescue, and relief forces (Fig. 2). The high operational capability and good equipment of the firefighting units enable them to be on standby 24 h a day to respond to natural and other disasters throughout the country. The system of volunteer fire departments in Slovenia is very well developed and represents an example of good practice for several countries in Europe and the world. The efficiency of the system is particularly evident during major natural disasters, where volunteer firefighters are indispensable. From the very beginning, firefighting was not only a service organized for emergencies to save people and their property, but it was also very important from a social and cultural point of view. Since the beginning of Slovenia's statehood, people were connected to each other through firefighting organizations, where national identity and cultural unity were built (Kramer Stajnko 2019). Slovenia today has about two million citizens, including more than 165,000 members of volunteer firefighting organizations and more than 50,000 operational firefighters ready to intervene at any time in the event of an accident. Currently, Slovenia has 1,300 volunteer fire departments and forty-five volunteer industrial fire departments affiliated with 120 firefighting associations. The entire country is divided into seventeen firefighting regions, which are an informal form of association. On the national level all associations are joined under The Firefighting Association of Slovenia (Firefighting Association of Slovenia 2020). The system in Slovenia is unique and recognized worldwide, with its high percentage of volunteer members and its high operational capability

.

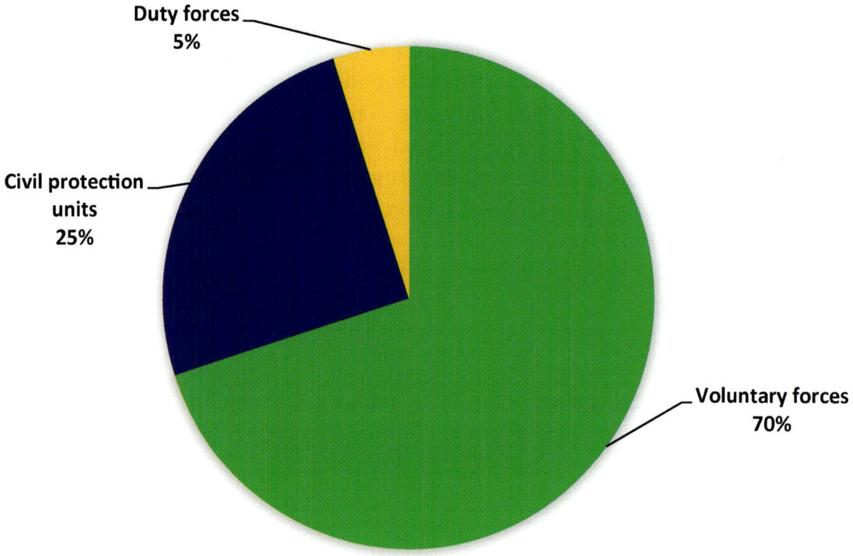

Fig. 2 Share of voluntary forces, civil protection units and duty forces among protection, rescue, and relief forces (ACPDR 2022)

4.1 *Activities, Education, and Training of Volunteer Firefighters*

For effective emergency response, it is crucial that firefighters are educated and trained. The accidents in which firefighters intervene in Slovenia are not only fires, as was the case in the past. The scope of tasks has expanded to include natural disasters (floods, storms, etc.), traffic accidents, technical, and other disasters (Fig. 3). In the curricula for training of volunteer firefighters, there is a wide range of technical content in which they learn how to use the equipment and gain basic knowledge of the hazards they can expect to encounter during operations. In addition, basic knowledge of organization, regulation, information technology, pedagogy, and andragogy are required to manage a fire department and are also included in the curriculum. The training system of volunteer fire departments is adapted to both the available time and the age structure of the members. It is assumed that there are young people in the firefighting organization who need basic skills, adult people who are active and need to improve specific skills, and older members who can no longer go on missions but also play an important role and are involved in other activities of the organization. The organization offers each of these groups the opportunity to acquire additional skills tailored to their abilities and the nature of their work. The main goal of training for volunteer firefighters is to acquire practical knowledge that they will need during operations. Training for volunteer firefighters in Slovenia can be divided into three groups (Pravila gasilske službe prostovoljnih gasilcev 2016): basic training, additional training, and permanent training. Volunteer firefighters are trained according to a valid curriculum prescribed by the Minister of Defense in accordance with the Fire Service Act (2005).

4.2 *Basic Training*

Volunteer firefighters acquire basic operational and organizational skills needed for their work in various fields through basic training. After each successfully completed course, they receive a rank that represents their level of training in the firefighting organization. Depending on the rank they have achieved, they can take on various tasks in the organization or during operations. In Slovenia, there are seven levels of basic training for volunteer firefighters (Pravila gasilske službe prostovoljnih gasilcev 2016): trainee firefighter course, firefighter course, senior firefighter course, junior firefighter course, captain fire officer course, chief fire officer course, and senior fire officer course. In Table 1 an overview of the basic training courses with duration and skills is presented. The courses from trainee firefighter to junior firefighter are organized at the municipal level, the course for captain fire officer is organized at the regional level, and the courses for chief fire officer and senior fire officer are organized at the national level—i.e., the Firefighting Association of Slovenia.

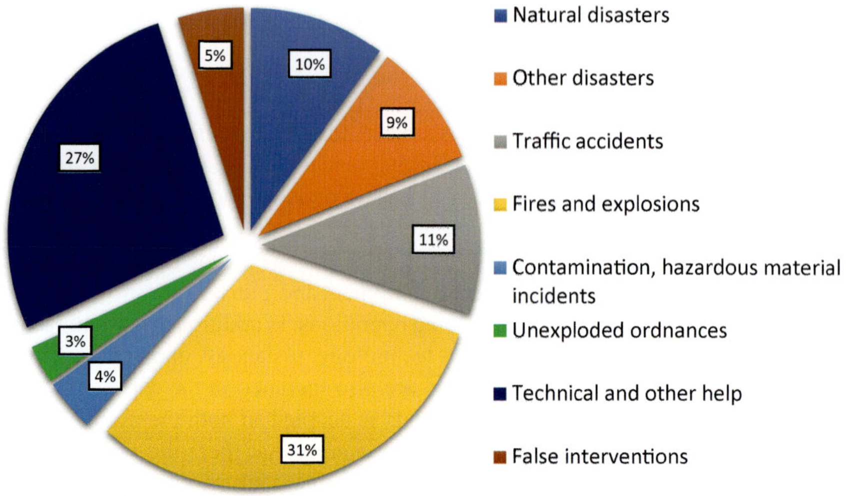

Figure 3: An overview of the type and number of accidents in which firefighters had to intervene in

Fig. 3 An overview of the type and number of accidents in which firefighters had to intervene in 2020 and 2021 (ACPDR 2022)

4.3 Supplementary Training

The main purpose of the additional training is for firefighters to acquire special knowledge they need in various operational situations, such as traffic accidents, respiratory protection, indoor firefighting, and extinguishing. In this case, the courses focus on practical work, while theory is reduced to a minimum. Firefighters have the opportunity to attend the courses in the following areas: machinist, respiratory protection, attack and extinguishing of indoor fires, technical rescue training, hazardous materials, outdoor fires, rescue from water, training for judges at competitions, training, first aid, informatics in the organization of firefighting, leadership of women, leadership of young people, and prevention.

5 The COVID-19 Pandemic in Slovenia

On March 4, 2020, Slovenia announced its first case of COVID-19; a state emergency was implemented on March 12, 2020, with the total lockdown of the whole country. The first recorded COVID-19 fatality was on March 14, 2020. As of February 2, 2022, confirmed cases reached 882,364 with 805,332 recoveries and 6,236 total deaths. Slovenia has 2.0 million people and an average of 293.4 deaths/million, and it was ranked eleventh for the most losses due to COVID-19, globally. The COVID-19 pandemic evolution in Slovenia can be divided into five waves: the first in the spring

Table 1 Basic training courses for volunteer firefighters in Slovenia

Course	Condition for the participation	Duration in hours		Gained skills	Trained for
		Theory	Training		
Trainee firefighter	Every member who turned 16 years	29	5	Fundamentals of volunteer, firefighting organization, fire prevention, first aid, protection against natural and other disasters	Participation on the interventions only under mentor supervision, less demanding tasks
Firefighter	Already gained rank of trainee firefighter	57	92	All theoretical and especially practical knowledge necessary for efficient and safe work on interventions	Individual work or for work in leaded group
Senior firefighter	Already gained rank for firefighter for at least 2 years	47	45	Skills to lead a group of firefighters (2–3 firefighters)	Leading a group of 2–3 firefighters and dividing the work among them
Junior fire officer	Already gained rank for senior firefighter for at least 2 years, finished vocational school	69	44	Skills to lead a fire department (up to 10 firefighters)	Leading a group of up to 10 firefighters and divide the work among them
Captain fire officer	Already gained rank of junior fire officer at least 2 years, finished high school	63	35	Skills to lead a fire brigade (up to 18 firefighters)	Recognizing hazards and dividing the work among the firefighters
Chief fire officer	Already gained rank of fire officer at least 2 years, finished college	53	26	Skills to lead a fire company (up to 26 firefighters)	Recognizing hazards and dividing the work among the fire departments
Senior fire officer	Already gained rank of chief fire officer, member of the authorities of Firefighting association of Slovenia	16	4	Research of current content and issues within the fire department organization	Development of the organization of the firefighting system

of 2020, the second in the fall and winter of 2020, the third in the spring of 2021, the fourth during the fall 2021 (delta), and the fifth in winter 2021/2022 (omicron). Several measures were applied by the Slovenian government to limit the spread of the virus and to prevent the collapse of the health system, e.g., total lockdown of the country (i.e., three longer periods)—which stopped all public life with restricting movement and gatherings and closing of kindergartens and schools, curfew, and mandatory quarantine for all infected and those who have been at high-risk contact. However, the health system in Slovenia has endured the crisis, but several problems with the system came to the surface. A detailed course of the COVID-19 pandemic is presented Table 2.

6 Impact of the COVID-19 Pandemic on the Activities of the Volunteer Firefighting System in Slovenia

To determine the impact of the COVID-19 pandemic on the activities of volunteer fire departments, a comprehensive study was conducted to address four questions.

- What impact did the reduced training and education activities have on the operations of the operational units in the volunteer fire departments?
- How were the operational capabilities of volunteer fire departments affected during the pandemic?
- How local communities and the state responded to the needs of fire units during the pandemic?
- What actions need to be taken immediately to create the conditions necessary to organize volunteer fire departments in the future?

The impact of reduced training and education activities is summarized from the official statistics of the Firefighting Association of Slovenia. In addition, the survey was conducted within the Slovenian firefighting departments based on an online questionnaire from December 9–28, 2021, i.e., during the fifth wave of COVID-19. The non-probability sampling technique was used; the survey was sent to all 1,300 volunteer firefighters in Slovenia, of which 437 members responded. The questionnaire contained sixteen questions of different types, e.g., dichotomous questions (yes/no), open questions, and questions with a rating scale, divided into two sections. The first section contained questions about the impact of the pandemic on the operational activities of the organization, e.g., whether they had a problem ensuring ready firefighters in case of an emergency, how many and what kind of permanent training was organized, and whether they used videoconferencing for training. The second section included questions about the social activities of the members during the COVID-19 pandemic, e.g., if they noticed an intrusion of members, if there was a problem with the integration of young people in the organization, how they substituted the activities they could not implement, such as meetings, general assemblies, etc. Finally, the

Table 2 Course of the COVID-19 Epidemic in Slovenia

Date	Measures
First wave (Spring 2020)	
March 4, 2020	The first detected case in Slovenia
March 7, 2020	Public indoor gatherings limited to fewer than 500 people
March 9, 2020	Limited outdoor gatherings per 500 people, indoor gatherings per 100 people
March 12, 2020	Epidemic declared
March 14, 2020	First death from a coronavirus in Slovenia
March 16, 2020	All educational institutions closed, public transport suspended
March 16, 2020	All stores except food and pharmacies closed
March 17, 2020	Prohibition of air traffic
March 19, 2020	Prohibition of public gatherings, disinfection of buildings with multiple housing units
March 30, 2020	Prohibition of movement outside the municipality
April 28, 2020	Some restrictions lifted (libraries, museums, shops)
May 4, 2020	Allowed outdoor areas of restaurants and bars
May 5, 2020	Allowed movement between communities
May 11, 2020	Public passenger transport starts
May 14, 2020	The government cancels the epidemic
May 31, 2020	End of the epidemic (wave 1)
Second wave (Autumn 2020)	
September 4, 2020	Mandatory wearing of masks in enclosed public spaces
September 15, 2020	Mandatory wearing of masks outdoors
October 16, 2020	Restrictions on regions, prohibited outdoor and private gatherings of more than ten people, prohibited events and religious ceremonies
October 17, 2020	Closed restaurants, distance learning
October 19, 2020	An epidemic has been declared again
October 20, 2020	Permitted sports activity at higher levels and members of the same household
October 20, 2020	Police hours between 9pm and 6am
October 27, 2020	Restriction of movement to municipalities, closed non-essential goods stores
December 13–23, 2020	Temporary release (shops, public transport), allowed crossing of municipal borders

(continued)

Table 2 (continued)

Date	Measures
December 24–25, 2020, and December 31, 2020-January 1, 2021	Mitigation of measures—socializing of six persons
December 26, 2020	The first vaccine in Slovenia
January 5, 2021	Schools with a customized program are opened
January 25, 2021	Schools for first triad students are opened, loosening restrictions across the country
Third wave (2021)	
month of March 2021	Re-increase in the number of infections
March 28, 2021	Shortening of police hour 10pm–5pm
April 1–11, 2021	Complete closure of the country, distance

questionnaire collected some ideas on what actions should be taken by the decision makers to restore the original state.

6.1 Impact of Reduced Education Activities

Educational activities were measured by the number of basic and supplementary training courses organized and the number of participants in these courses. Table 3 shows the number of organized courses and participants in the basic training courses over the past five years, and Fig. 4 provides a graphical representation of participants in each course. A significant decrease in courses and participants can be observed in 2020, when the measures due to the COVID-19 pandemic were the strongest. However, senior firefighter and junior firefighter officer courses have seen the largest decrease. In the future, it will be necessary to motivate volunteer firefighters to attend the courses intensively to maintain personnel continuity in the volunteer firefighting units and their operational capability in case of interventions.

In the field of supplementary training, we highlight the three most important and most frequently organized supplementary training courses to determine the impact of the COVID-19 pandemic, such as respiratory protection, attacking and extinguishing indoor fires, and technical rescue training. The number of organized courses and participants in the last five years are shown in Tables 4 and 5, and the graphical representation of the data can also be seen in Fig. 5. In 2020 and 2021, there is a significant decrease in the number of participants, especially in the fields of respiratory protection training and technical rescue, as the measures due to the COVID-19 pandemic prevented the organization of training courses. To ensure fully operational firefighting units in the future, special attention must be paid to the education and training of volunteer firefighters.

Table 3 Number of organized courses and participants of basic training in last five years

Year	2017		2018		2019		2020		2021	
Course	Nr. of courses	Nr. of participants	Nr. of courses	Nr. of participants	Nr. of courses	Nr. of participants	Nr. of courses	Nr. of participants	Nr. of courses	Nr. of participants
Trainee firefighter	114	2335	122	2383	141	3001	53	1060	96	1917
Firefighter	27	449	58	1034	66	1258	36	570	26	492
Senior firefighter	30	510	30	601	41	732	9	172	7	129
Junior fire officer	13	223	17	367	13	231	8	109	5	43
Captain fire officer	11	250	2	52	6	141	0	0	1	30
Senior fire officer	1	26	1	29	1	23	1	27	1	25

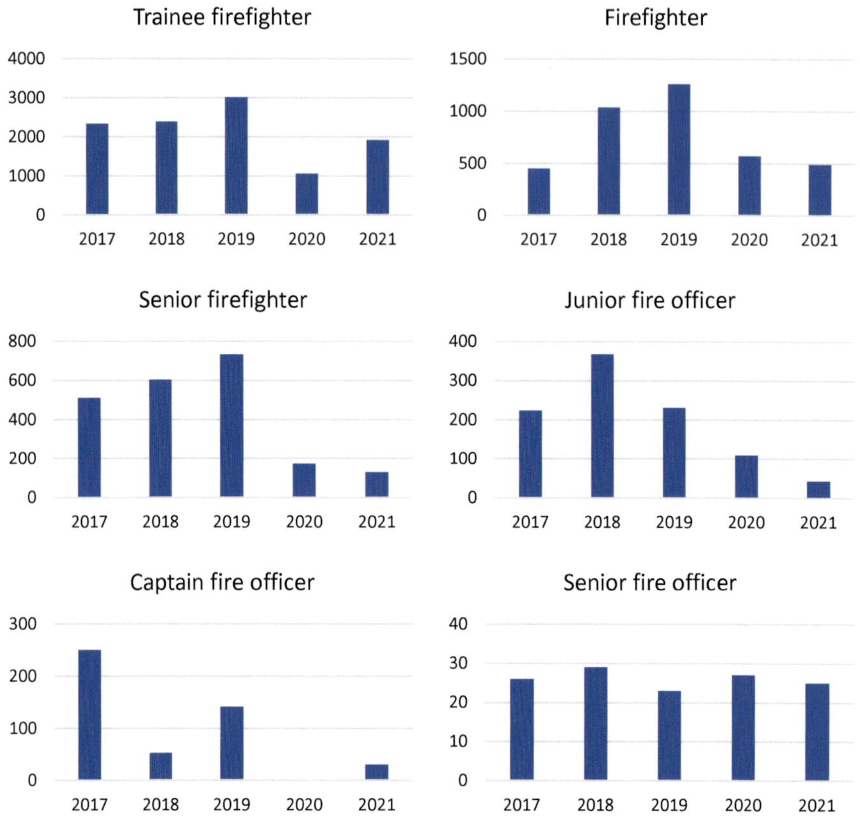

Fig. 4 Number of volunteer firefighters on basic training in last five years

To determine the relationship between reduced educational activities and the operational capability of firefighting units, data on volunteer firefighters work and participation in operations over the past five years are provided in Table 5. It is more than obvious that volunteer firefighters represent the majority of all participants in operations. However, in 2020 and 2021, no decrease in participation of volunteer firefighters was observed, which means that the operational capability is not yet in jeopardy. However, it is too early to determine the impact of reduced training activities in 2020 and 2021, as this deficit will only become apparent in the long-term period.

Table 4 Number of organized courses and participants of supplementary training in last five years

Year	2017		2018		2019		2020		2021	
Course	Nr. of courses	Nr. of participants	Nr. of courses	Nr. of participants	Nr. of courses	Nr. of participants	Nr. of courses	Nr. of participants	Nr. of courses	Nr. of participants
Respiratory protection training	37	799	31	664	33	702	16	335	30	404
Attack and extinguishing of indoor fires training	52	976	49	945	50	975	30	574	59	888
Technical rescue training	17	451	16	429	16	423	7	170	7	101

Table 5 Number of interventions and participants of volunteer firefighting units in the past five years

	2017	2018	2019	2020	2021
Nr. of all interventions	25,652	19,578	20,472	23,488	20,199
Nr. of all interventions where volunteer fire units participated	15,800	10,838	12,139	14,845	11,955
Nr. of all participants on interventions	517,006	230,798	158,027	174,225	151,376
Nr. of volunteer firefighters	470,415	183,704	116,247	130,465	108,834

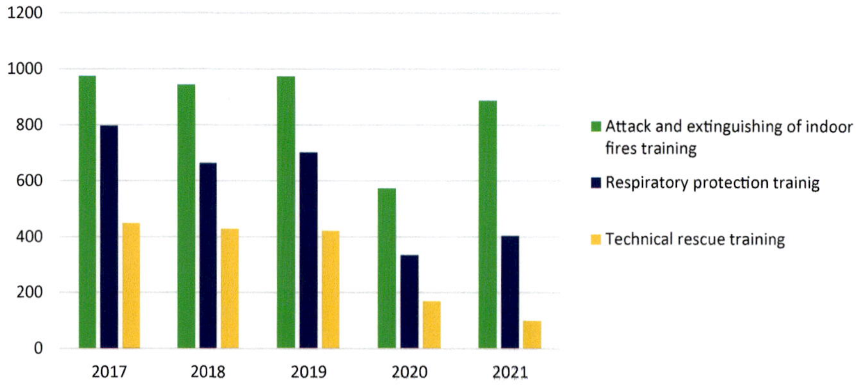

Fig. 5 Number of volunteer firefighters on selected supplementary training in last five years

6.2 *Impact of the Pandemic on the Organization's Operational and Social Activities*

In general, the results of the survey show that the COVID-19 pandemic has affected the activities of volunteer fire departments, especially in terms of social activities and youth, but there is still no impact in the field of operational work, as all missions have been successfully completed. The results of the survey show that in 84% of all responses, organizations did not experience major problems in ensuring appropriate operational firefighting units in case of emergency, e.g., due to quarantine, illness, fear of contagion. Problems were noted in 16% of cases, with the most frequently cited problems being fear of infection, illness, and quarantine. The most frequently collected responses on the question of what special measures they took during the COVID-19 pandemic, are presented in Fig. 6.

In addition, 81% of the responses indicate that fire units had sufficient protective equipment when responding to operations during the pandemic. The equipment came from various sources, as can be seen in Fig. 7.

When asked if they followed the instructions of the Firefighting Association of Slovenia, 93% answered yes, which shows that the lower levels of the organization have high expectations of the federal association. In addition, the question about

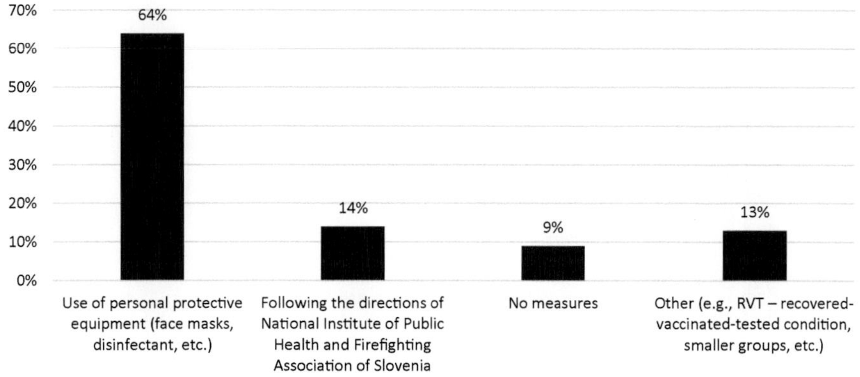

Fig. 6 Measures in fire departments during the COVID-19 pandemic

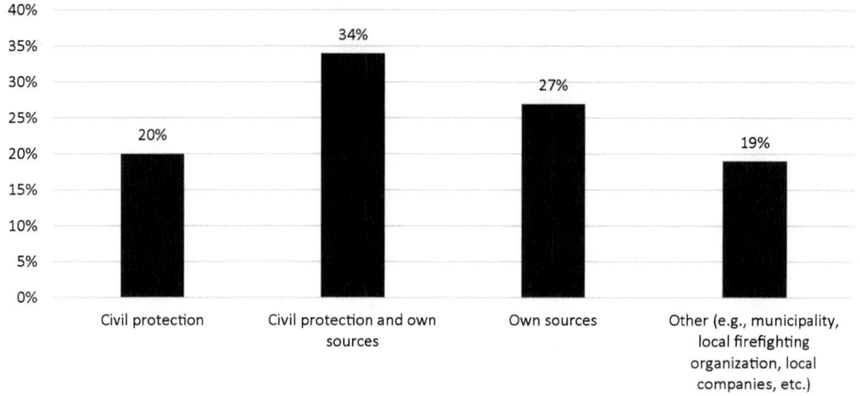

Fig. 7 Finding of the protective equipment in volunteer firefighting units during pandemic

how much practical training (as a part of permanent training) was organized within one year before the COVID-19 pandemic and then in the years 2020 and 2021. The results are shown in Fig. 8. The frequency of the number of organized practical training groups is divided into four, where the first group includes 0–10 training sessions in the period of one year, the second 10–20, the third 20–40, and the last more than forty. It should be emphasized that before 2020 about 2% of organizations did not organize practical training sessions, in 2020 the percentage was much higher, namely more than 17% of departments did not organize any practical training, in 2021 the conditions were better, namely only 6% of departments did not organize any practical training. However, it is obvious that the number of practical training sessions decreased drastically due to the restrictions caused by the pandemic, but still firefighters made efforts to organize basic practical training.

During the pandemic, some activities (i.e., theoretical parts of courses, work, and other meetings) were conducted via videoconferencing systems. The survey

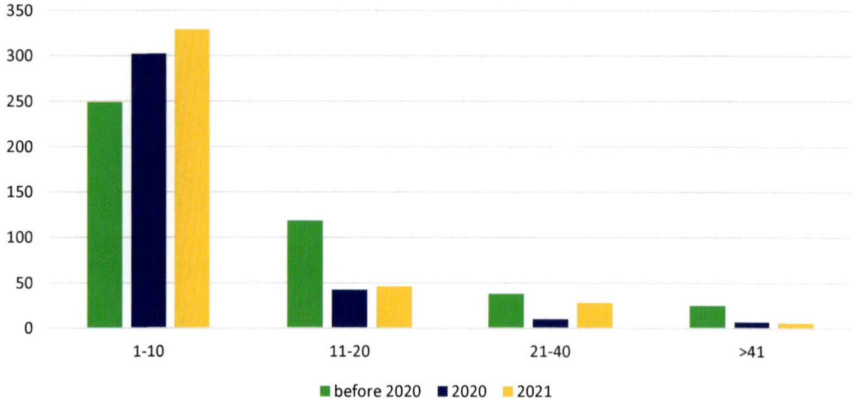

Fig. 8 Number of organized practical training before the year 2020, and later in the year 2020 and 2021

showed that only 55% of all respondents used the various options for remote training and sessions. Several benefits of remote sessions were highlighted, such as safety from infection, sessions were shorter and more efficient, more participants had the opportunity to attend sessions or training because they could stay at home, and at least it offered a way to maintain contact with colleagues in the fire unit. The following were highlighted as disadvantages: no personal contacts, inefficient education due to the passivity of participants and fewer opportunities for open debate, technical problems (i.e., internet connection, software, etc.), the older population is not versed in the use of computers, and no opportunities to conduct practical training. Finally, respondents believe that the COVID-19 pandemic has left the greatest impact in the following areas:

- due to the lack of activities, there is a big problem in motivating young people to become members of a firefighting organization,
- lack of social contact among members,
- operational capability,
- fewer sources of income,
- no activities for women and elderly people, and
- fewer courses, training, and competitions.

Moreover, when asked if they have seen a drastic decline in membership of the organization, 79% of respondents said they have not observed such a problem. When asked if they have problems getting young people involved in the organization, 82% answered that it is a problem to motivate and attract young people to become volunteer firefighters.

7 Conclusions

Slovenian volunteer firefighting organization represents the most important and largest part of the system of protection, rescue, and relief forces. It is highly developed and ensures readiness in all emergencies. However, the COVID-19 pandemic and the measures taken had consequences for the activities of the volunteer fire departments that are not yet obvious but will have a long-term impact. Therefore, several measures must be taken immediately at the local and national level to mitigate the negative impact of the pandemic and ensure the operational capability of volunteer fire departments in the future. The following commendations outline the necessary measures to address the situation effectively.

It is urgent to immediately start recruiting children and young people as volunteer firefighters to ensure full operational capability in the future. Special attention should be paid to the training of volunteer firefighters. It is expected that the consequences of reduced training activities will be felt in a few years, so it is important to organize as much training as possible to ensure well-trained firefighters and firefighting units. The survey showed that firefighters from all over the country have great expectations from local and national authorities as well as from The Firefighting Association of Slovenia and were confused during the pandemic due to the many messages from all sides; it is important that The Firefighting Association of Slovenia takes care to provide clear instructions on the measures that apply to firefighters and their work. Moreover, volunteer firefighters should be treated as an exception, as they have to intervene and save lives in all kinds of accidents. Furthermore, it is the responsibility of national authorities to ensure that the volunteer firefighting system will be maintained and developed in the future. Some important areas concerning the status of volunteer firefighters should be addressed, such as the benefits of volunteer firefighters in the social security system, motivating and rewarding employers for employing volunteer firefighters. Finally, the system for financing and procuring equipment is well developed at both the local and national levels, ensuring a continuous supply of equipment to all firefighting units throughout the country. There is a lack of establishment of adequate facilities as there is still no funding for this purpose.

References

ACPDR (2022) Administration of the Republic of Slovenia for civil protection and disaster relief. http://www.sos112.si/eng/

Cirella GT (2020) Sustainable human-nature relations: environmental scholarship: economic evaluation, urban strategies. Springer, Berlin. https://doi.org/10.1007/978-981-15-3049-4

Cirella GT (2022) Human settlements: urbanization, smart sector development, and future outlook. Springer. https://doi.org/10.1007/978-981-16-4031-5

Cirella GT, Zerbe S (2014) Quizzical societies: a closer look at sustainability and principles of unlocking its measurability. Int J Sci Soc 5:29–45. https://doi.org/10.18848/1836-6236/CGP/v05i03/51427

Cirella GT, Wanjiku S, Paczoski A, Tiruneh S (2020) Human-nature relations: the unwanted filibuster. In Cirella GT (ed) Sustainable human–nature relations: environmental scholarship, economic evaluation, urban strategies. Springer, pp 3–22. https://doi.org/10.1007/978-981-15-3049-4_1

Cirella GT, Russo A, Benassi F, Czermanski E, Goncharuk A, Oniszczuk-Jastrząbek A (2021) Energy re-shift for an urbanizing world. Energies 14:5516. https://doi.org/10.3390/en14175516

Fire Service Act (2005) Official gazette of the Republic of Slovenia, No. 113/05 - official consolidated text, 23/19 and 189/20 – ZFRO

Firefighting Association of Slovenia (2020) VULKAN – information system of voluntary firefighting departments

Fu Y, Yang X, Wang T, Supriyadi A, Cirella GT (2022) Spatial pattern characteristics of the financial service industry: evidence from Nanjing, China. Appl Spat Anal Policy 15. https://doi.org/10.1007/s12061-021-09409-z

Jiao G, Lu L, Chen G, Huang Z, Cirella GT, Yang X (2021) Spatiotemporal characteristics and influencing factors of tourism revenue in the Yangtze River delta urban agglomeration region during 2001–2019. Sustainability 13:3658. https://doi.org/10.3390/su13073658

Kahraman C, Orobello C, Cirella GT (2022) Changing dynamics with COVID-19: future outlook. In: Cirella GT (ed) Human settlements: urbanization, smart sector development, and future outlook. Springer, pp 235–252. https://doi.org/10.1007/978-981-16-4031-5_13

Kladnik A (2020a) A nation of joiners. Volunteer firefighters and Slovenian nation- and state-building from below. Südosteuropa 68(2):148–175

Kladnik A (2020b) Volunteering and voluntary associations in the Post-Yugoslav states. An introduction. Südosteuropa 68(2):125–129

Kramer Stajnko J (2019) Organization and operation of voluntary firefighting organizations in Slovenia. Ujma 33:250

Kramer Stajnko J, Petek F (2019) Impact of the structural flood protection measures on the activities of protection, rescue and relief forces. World construction forum, Ljubljana, Slovenia

Leskovec R, Kramer Stajnko J (2021) The training of Slovenian volunteer firefighters. Ujma 34–35:425

Lewandowski R, Cirella GT (2022) Performance management systems: trade-off between implementation and strategy development. Oper Manag Res 16. https://doi.org/10.1007/s12063-022-00305-4

Oniszczuk-Jastrząbek A, Czermanski E, Cirella GT (2020) Sustainable supply chain of enterprises: value analysis. Sustainability 12:su12010419. https://doi.org/10.3390/su12010419

Pravila gasilske službe prostovoljnih gasilcev (2016) Gasilska zveza Slovenije, Ljubljana

Ratchev V (2015) Slovenia: capabilities, organizations, policies, and legislation in crisis management and disaster response. Institute of Information and Communication Technologies CSDM, Centre for Security and Defense Management IT4Sec Reports 123

Ratej M (2013) Drugi jugoslovanski gasilski kongres v Ljubljani – zadnja velika manifestacija katoliške politične misli pred začetkom druge svetovne vojne. Časopis Za Zgodovino in Narodopisje 84(4):71–87

Russo A, Cirella GT (2020) Urban sustainability: integrating ecology in city design and planning, in: Cirella GT (ed) Sustainable human–nature relations: environmental scholarship, economic evaluation, urban strategies, advances in 21st century human settlements, pp 187–204. Springer, Singapore. https://doi.org/10.1007/978-981-15-3049-4_10

Spring C, Cirella GT (2022) Fostering sustainable development: green energy policy in the European Union and the United States. In: Human settlements: urbanization, smart sector development, and future outlook. Springer, pp 101–137. https://doi.org/10.1007/978-981-16-4031-5_7

The Role of Urbanization on Temperature and Precipitation in Africa

Bruno E. Ongo Nkoa, Jacques S. Song, and Giuseppe T. Cirella

Abstract Urbanization has profoundly affected the environment and climate on a global scale which has led to intensified weather changes around the world. The aim of this chapter is to examine the effects of urbanization in Africa on climate change using a holistic approach known as the dynamic's technique. Temperature and precipitation are the two dimensions of climate change measured to show a correlation between climate change and urbanization on the African continent. Using a sample of 52 African countries, the research estimates a panel data model using the ordinary least squares and the general least squares methods over the period 1991–2020. The results show preliminary conclusions that regardless of the dimension considered, urbanization increases climate change in Africa. The research was conducted using robustly formulated datasets, encompassing urban population growth as an indicator of urbanization and sensitivity calculations to test agglomeration urban population, urban area, and urbanization rates.

Keywords Climate correlation · Ordinary least squares · Generalized least squares · Dynamic's technique · Urban population

B. E. Ongo Nkoa (✉)
Center for Studies and Research in Economics and Management, University of Yaounde 2, Yaounde, Cameroon
e-mail: ongoema@yahoo.fr

J. S. Song
Faculty of Economics and Management, University of Ebolowa, Ebolowa, Cameroon
e-mail: jacquessimonsong@gmail.com

G. T. Cirella
Faculty of Economics, University of Gdansk, Sopot, Poland
e-mail: gt.cirella@ug.edu.pl

G. T. Cirella and B. Dahiya (eds.), *City Responses to Disruptions in 2020*, Advances in 21st Century Human Settlements, https://doi.org/10.1007/978-981-99-7988-2_11

1 Understanding Global Urbanization Trends

Ongoing climate change consequently requires a revolution in the conception of medium- and long-term urban planning; it is no longer a question of "simply" qualifying the environmental impacts of urban or development projects, but also considering the impacts of environmental changes on planned operations. With appropriate governance structures, cities can be places of innovation and efficiency that reduce the causes of climate change and effectively protect themselves from its effects. This can be achieved with alternative modes of transport, high-performance buildings, revegetation, alternative rainwater management, densification, and functional diversity. Land use planning, in general, is one of the keys needed to be grasped in organizing a society that is resilient to the bearings of climate change in the long-term, and to prevent the risks of conflicts related to access to water and energy. Cities are places of economic, social, and cultural development par excellence. All forms of this development have been evolving at an unprecedented pace since the end of the twentieth century, inducing radical and irreversible transformations in urban environments. Currently, more than 70% of the world's population lives in cities, compared to only 15% in 1900 and 50% in 2007. This galloping expansion of urbanization fully contributes to the development of industries and services, while promoting qualitative improvement and unequivocally quantitative measure of the level and quality of life of populations (Cirella and Zerbe 2014a; Cirella 2020; Russo and Cirella 2020; Cirella et al. 2020, 2021; Kahraman et al. 2022; Spring and Cirella 2022; Fu et al. 2022). Nonetheless, it is on the environmental level, including its climatic component, that the impacts of the acceleration of the urban phenomenon are felt in a negative way (Oniszczuk-Jastrząbek et al. 2020; Cirella 2022). The urban environment, due to the artificialization it imposes on urban space, is at the origin of radiative, thermal, dynamic, and water processes which modify the thermodynamic exchanges between the surface and the atmosphere. Thus, and depending on the size and density of the city, this artificialization results in the creation of a microclimate.

In fact, it can be argued that rapid urbanization can make the world's population more vulnerable to the effects of climate change. More than half of the world's population is now urban, and by 2050 an additional 2.5 billion people are expected to live in cities (Moore et al. 2003; Parnell and Robinson 2012; Dumont 2018). Torrential rains and storms are becoming more frequent in densely populated cities. These phenomena affect people living in marginalized and informal settlements, such as slums, are hardest hit. Desertification destroys the arable land needed to feed growing urban populations. In addition, rising sea levels threaten the inhabitants of coastal regions, deltas, and small island states. Precisely in urban areas, the effects of urbanization are becoming increasingly complex because of the interactions between human activities and environmental conditions (Russo and Cirella 2017). The impact of urbanization on climate change is derived from changes in atmospheric heat, dynamic conditions, and composition in urban environments. These changes are largely owing to urban heat island effects, underlying surface changes, and aerosol emissions (Du et al. 2016). Indeed, like Asia, Africa is urbanizing at twice the rate of Europe. Africa, for its part, has experienced the same evolution in 60 years, i.e., almost in half the time. This urbanization increased by 5.9% between 2000 and 2015. In 2015, nearly 472 million Africans lived in urban areas. In 20 years, the urban population has almost doubled in Africa, from 237 million in 1995 to 472 million in 2015. Africa's urban population is expected to almost double again between 2015 and 2035. Not so long ago, in 1990, Africa was the region of the world with the fewest city dwellers (i.e., 197 million). Ironically, in 2020, it will probably have the second largest urban population in the world (i.e., 560 million) after Asia (i.e., 2.3 billion).

Africa's unprecedented urbanization is favored by the demographic change carried by the young population in search of a better life in the cities (Fig. 1). This rapid evolution explains the imbalances of development between urban and rural areas. Population growth influences socioeconomic, environmental, and political organization. Public authorities should harness human and technological resources to improve productivity, mobility, and human interactions, to master the disruptions that accompany urbanization (Abebe et al. 2019). Urban growth, unaccompanied by adequate policies, induces a deepening of inequalities and fragilities. Today, more than half of the world's population lives in urban areas. Even if they only cover 2% of the earth's surface, cities thus concentrate the majority of GHG emissions from the residential, tertiary, industrial, and transport sectors. Population growth will amplify this phenomenon of urban concentration, i.e., cities will expand, and with them the artificialization of soils, weakening biodiversity and ecosystems. At the same time, these urban areas are highly vulnerable to climate change, rising sea levels, and extreme weather events (e.g., droughts, floods, and storms) that will impact infrastructure and deteriorate access to basic networks and services (e.g., water and energy).

Contemporarily, to reduce the challenges associated with climate change, a new concept of cities is required, particularly of urbanization. Some challenges are reflected in the United Nations Sustainable Development Goals (SDGs), in particular, SDG 11, (i.e., sustainable cities and communities) and SDG 13 (i.e., measures relating to the fight against climate change). Urbanization, despite the risks, offers opportunities to mitigate climate change. Considering that environmental, economic, political, and social factors all play critical roles in the capacity of communities to adapt to climate change, it has become increasingly relevant to consider how urbanization affects adaptive capacity of countries. The effects of urbanization and climate change are converging in dangerous ways. Cities large and small are already the hardest hit by natural disasters, such as floods and tropical storms. Many of the world's largest cities located along coasts, rivers, and in floodplains are the most vulnerable to natural disasters. According to the State of Africa's Cities report, by 2050, 200 million Africans could be displaced due to the effects of climate change, which will put a heavy strain on the capacities of cities and their resources (UN-Habitat 2018).

As an extension of this theoretical framework and inspired by empirical methodologies, i.e., widely documented in the literature, the objective of this chapter, which constitutes moreover its originality, is to examine the effects of urbanization on climate change throughout the African continent. For this, we consider a sample of 52 African countries over the period 1991–2020 and applied the methods of

Fig. 1 Examples of urbanization in selected African cities: (*top*) Windhoek, Namibia—9 July 2018; (*middle left*) Dar es Salaam, Tanzania—11 May 2020; (*middle right*) the Ben Schoeman Freeway connecting Johannesburg and Pretoria in South Africa—22 August 2018; and (*bottom*) urban sprawl outside of Maseru, Lesotho—3 September 2018. *Source* (*top*, *middle right*, and *bottom*) Photographs by Giuseppe T. Cirella; (*middle left*) photograph by Peter Mitchell on Unsplash, Creative Commons Public Domain

ordinary least squares (OLS) and generalized least squares (GLS). The preliminary findings reveal, overall, that urbanization increases climate change in Africa. For this purpose, and more specifically, this chapter contributes to the literature on the subject in three very important ways. First, it integrates, in addition to traditional determinants, urbanization as a relevant indicator of climate change. Second, it empirically completes and enriches the work of UN-Habitat (2018) which focused more on the theoretical description of urbanization as a factor accentuating situations of climate change. Third, it shows its polysemous nature, including the consideration that temperature and precipitation make it possible to grasp their complexity, to address relevant recommendations. This chapter is organized into a further four sections. The first section presents a selective review of the state-of-the-art in a purely theoretical manner. The second section succinctly outlines the different stages of the empirical strategy adopted. The third section discusses the results of the study, and the fourth focuses on suggested recommendations in terms of economic policy.

2 Effects of Urbanization on Climate Change

This section outlines the analytical framework useful for understanding the theoretical anchoring of climate change and draws up an empirical synthesis of the urbanization effects on climate change. The empirical work synthesized is inspired by previously analyzed theories.

2.1 Theoretical Synthesis

There are several theories that explain the climate change phenomena including the Milankovitch Theory which describes the relationship between the sun and the earth, and the Astronomical Theory. However, the theoretical anchoring of the effects of urbanization on climate change is explained by three theories. First, the theory of modernization (Hawley 1944) which maintains that urbanization occurs with the advancement of technology and the transition from the primary sector to the secondary and tertiary sectors, inducing damage to the environment through noise and odor nuisances that influence temperatures and rainfall. The second model, the urban bias theory (Bradshaw 1987) which emphasizes that urbanization and population growth are intimately linked to environmental degradation. It causes pollution of water, air, and soil by transport or industries, as well as the accumulation in the atmosphere of GHGs which cause climate change. Finally, the dependency theory (Kentor 1981) points out that the urbanization of developing countries is controlled by developed countries, the main culprits of atmospheric pollution and the accentuated degradation of the ozone layer. These theories have favored the emergence of two types of climate change. On the one hand, past climate change is characterized by a series of climate alterations marked by cold periods and warm periods. On the other hand, the current climate change is characterized by an increase in the average global temperature, as well as in the level of sea water. This phenomenon is most likely due to GHGs, such as CO_2, which are released into the atmosphere during the combustion of fossil fuels such as coal, oil, and gas, as well as during land use, e.g., with the deforestation of tropical forests. The literature outlines the issues, constraints, and relational understanding of the adoption of urbanization in the process of climate change.

The seminal paper of Barrios et al. (2006) examines how climate change and urbanization in Africa are related. They find that, an increase in the national urban share of 0.45% with a reduction in national rainfall of 1%. Climate change remains largely driven by the demand for urban habitat (Todaro 1969). Although the effect of urbanization on climate change has recently emerged, metaphors for futuristic and idealistic visions have circumscribed the concept of urban development. The emergence of the urban economy, dating back to the 1990s, promoted the perception of urbanization as a determinant for the human and sustainable development of countries. Such determinants have been linked to the decolonization of Africa which also, simultaneously, corresponds with the "lifting of legislation prohibiting the free internal movement of native Africans" (Barrios et al. 2006). A current map of Africa's political boundaries is presented in Fig. 2.

Fig. 2 Political map of Africa, 2022. *Source* Wikipedia (2022), Creative Commons Public Domain

2.2 Empirical Synthesis

The empirical anchoring effects of urbanization on climate change date back to the seminal work of Wigley and Farmer (1982), Colacino and Rovelli (2016), and Kukla et al. (1986) whose focus of research was on urbanization in tropical cities. The tropical cities examined by Wigley and Farmer (1982) did not concentrate on major cities of the developing world, where population booms and large human fluxes into cities have created disorganized settlements causing cities to grow in a fast and chaotic manner. Previous studies have investigated how urbanization affects the intensity, amount, and spatial patterns of precipitation in and around urban areas based on empirical measurements and numerical model simulations. Such research has noted that urbanization may cause a precipitation increase upwind or downwind (Changnon 1979; Lin et al. 2011; Zhang et al. 2017c; Yu et al. 2020). Therefore, the impact of urbanization on moderate and heavy convective precipitation is particularly significant (Cirella et al. 2014; Kusaka et al. 2014; Zhang et al. 2017c). The influence of urbanization on atmospheric properties has been theoretical investigated in a vast number of cities around the world (Oke 1979; Landsberg 1981; Cayan and Douglas 1984; Goodridge 1987). Thus, a few studies investigated urban heat island intensity in a short time period (Park 1986). But these results are difficult to link to seasonal and annual mean temperature. Still, some studies have used temperature–time series of sufficient length to demonstrate the positive temperature trend in urban cities (Yagüe et al. 1991; Jauregui et al. 1992; Karaca and Asian 1993). Nonetheless, these studies have underlined the fact that urban areas are warmer than their surroundings, especially during clear, calm nights and this is normally a result of the delayed cooling of the city compared to its surrounding rural sites.

Several studies have highlighted that urbanization (and different forms of modern activities) is the reason for the increase in climate change (Lin et al. 2011; Zhang et al. 2017a, b; Yu et al. 2020; Czermanski et al. 2021). These studies found urbanization increases precipitation upwind or downwind (Gero et al. 2006; Kusaka et al. 2014; Yan et al. 2020). Thus, it can be argued that urbanization increases climate change through moderate and heavy convective precipitation. Consequently, Wang et al. (2014) found that the increase in total annual precipitation in the Pearl River Delta is due mainly to an increase in rainstorms and extreme precipitation events under the effect of urbanization. Moreover, the impact of urbanization on the type of extreme precipitation has also been examined. According to Li et al. (2011), the urban effect of northern Taiwan seems to significantly enhance convective rainfall rather than stratiform rainfall, while Wen et al. (2020) found that urbanization increases frontal rainfall. Numerical simulation and observation studies have also revealed three major mechanisms affecting precipitation: urban heat islands (Wang et al. 2011; Lin et al. 2011; Zhang et al. 2017c), surface roughness changes caused by urban modification (Miao et al. 2011), and aerosol emission (Wang et al. 2011; Xiao et al. 2014; Veremchuk et al. 2016). Karaca et al. (1995) found that urbanization and industrialization in the southern part of Istanbul have a negative effect on regional cooling. Bargaoui and Nouri (2017) performed a panel data analysis of climate change approximated by the CO_2 emissions driving forces in the context of a panel of countries for the period 1890–2010 and found positive but statistically insignificant impacts of urbanization on CO_2 emissions. Zhang et al. (2017a, b, c) found evidence of a non-linear relationship between urbanization and CO_2 emissions in the sample of a panel of 141 countries over the period 1961–2011. According to the authors of this article, urbanization increases, leads to a fall in the level of CO_2, and at a certain point in time the relationship gets reversed whereby further urbanization accounts for higher levels of CO_2 emissions.

Dash and Mallick (2017), using annual data from 1990 to 2012 and an OLS estimation technique, found a positive linkage between urbanization and annual average temperature across the Middle East. Muntasir and Syed (2018) using Augmented Dickey-Fuller test, Johansen Cointegration test, and Vector Error-Correction Model causality test, and Granger causality test found that urbanization has a causal effect on climate change, notably on the GHG emissions and temperature change in the long run in Bangladesh. According to Yan et al. (2020) and Yang et al. (2014), rapid urbanization led to an increased frequency of disasters such as extreme rainfall events in urbanized areas. Many perspectives in these studies emphasize urbanization as a driver of climate change exposure. Recently, Li et al. (2021) considered two comparative urban development scenarios (i.e., 2000 and 2015) designed to examine the sensitivity of extreme precipitation to urbanization in numerical experiments. They found that urbanization significantly impacts climate change—precisely extreme precipitation and typhoon extreme rainfall.

From the non-exhaustive synthesis of the state-of-the-art, three observations emerge. Firstly, the work focuses more on investigating the direct effects of urbanization than the quality of the environment. Secondly, and to the best of our knowledge, studies have not focused on the effects of urbanization on climate change. Thirdly, the little consensual conclusions of empirical work justify new investigations in a context of urban development and proven climate change in Africa.

3 Building an Empirical Strategy to Assess the Role of Urbanization on Climate Change in Africa

In this section the different stages of the empirical strategy are adopted. First, the basis of the empirical model and its various specifications are detailed. Second, the different technical specifications from the estimates are used. Third, the endogeneity treatment process is reviewed. Lastly, the sample, time horizon, and data are introduced.

3.1 Empirical Model

The empirical model spurs from the work of Yan et al. (2020) and Yang et al. (2014) who both have adopted specifications considering climate change from Barrios et al. (2006). Its choice is justified by the fact that climate change cannot be explained by traditional models, but by a model of urban structure which would predict, that improvements in favor of urbanization should mitigate the magnitude of climate change. The reduced specification utilizes Eq. 1.

$$\mathrm{CC}_{it} = \alpha + \beta U_{it} + \lambda X_{it} + v_i + \mu_t + \varepsilon_{it} \tag{1}$$

where: CC_{it} = the dependent variable that describes climate change of country i at time t; U_{it}= represents urbanization of country i at time t; X_{it} = the vector of control variables that integrates foreign aid development, remittances, unemployment, trade openness, agricultural area, gross fixed capital formation, education, internet, and mobile phone; indices i and t provide information on countries and periods; v_i = captures unobserved country fixed effects; μ_t = reflects the temporal fixed effect common to all countries; and ε_{it} = the error term.

The description of the study variables is contained in Table 1.

3.2 Estimation Method

To determine the effects of urbanization on climate change, we used a dynamic model estimated by OLS, used as an initial framework in the analyses to give the general trend of the results according to Legendre (1805) and Gauss (1809). However, it minimizes the impact of measurement errors, does not consider the problem of heterogeneity, and assumes that the countries in the sample are perfectly homogeneous. Contrary to previous works that analyze the role of urbanization on climate change and either postulated a static analysis or a dynamic analysis, this chapter uses OLS to make a major contribution by not only considering the form of the link but also, and especially, the efficiency of the estimators retained. Moreover, we used GLS estimates. Two strong reasons plead for the choice of such a technique. The first is based on the nature of the data. By capturing the effect of alternation in power, the measurement of variables remains static and therefore not very mobile over time. The second is validated by the sample size, so narrowness does not pose major problems in the estimation process. Without claiming to be exhaustive, the use of this econometric technique provides at least three advantages (Davidson 2000): (1) it makes it possible to obtain, in the presence of lagged variables, estimators without bias, convergent and efficient; (2) it takes into account the unobservable factors that can have an impact on the attractiveness of countries' economic growth and trade openness (Rodrik 2014); and (3) it makes it possible to correct the simultaneity bias between the variables of

Table 1 Study variables for the model

Variable	Description	Source
Climate change	• Rainfall measured by the level of annual precipitation in mm • Temperatures measured by the annual temperature level in °C	Climate change knowledge (World Bank 2021) Portal (World Bank 2021)
Urbanization	• Urban growth: growth rate of urban population • Population in urban agglomerations: share of population in urban agglomerations as % of total population • Urban population: share of population in urban areas as % of population • Urban area: urban area in km^2	WDI (World Bank 2020)
Foreign public aid	• Development aid in value	WDI (World Bank 2020)
Remittances	• Share of earned income abroad	WDI (World Bank 2020)
Unemployment	• Number of unemployed persons aged 15–34 years, as a proportion of the total number of this population group	WDI (World Bank 2020)
Trade openness	• Development aid in value	WDI (World Bank 2020)
Agricultural area	• Area of agricultural land as % of total area	WDI (World Bank 2020)
Domestic investment	• Gross fixed capital formation relative to gross domestic product that captures private investment	WDI (World Bank 2020)
Education	• Number of students enrolled in secondary	WDI (World Bank 2020)
Internet	• Rate of Internet subscription of the population (i.e., % of the population per 100 people)	WDI (World Bank 2020)
Mobile phone	• Subscription rate of mobile phone usage per 100 people	WDI (World Bank 2020)

interest and control (Wooldridge 2010). However, the absence of autocorrelation of the residuals guarantees the obtaining of the estimators.

3.3 Data

The study covers 52 African countries (Table 2). The time horizon covers the period 1991–2020. However, for methodological reasons, the study period is divided into four to obtain nine sub-periods. The choice of period is dictated by the availability

Table 2 List of African countries utilized in the study

Subregion	Countries
Northern Africa	Algeria, Egypt, Libya, Morocco, Sudan, Tunisia
Eastern Africa	Burundi, Comoros, Djibouti, Ethiopia, Kenya, Malawi, Mauritius, Mozambique, Rwanda, Seychelles, Somalia, Tanzania, Uganda, Zambia, Zimbabwe
Middle Africa	Angola, Cameroon, Central African Republic, Chad, Congo-Democratic Republic of the, Congo-Republic of the, Equatorial Guinea, Gabon, São Tomé and Príncipe
Southern Africa	Botswana, Eswatini, Lesotho, Madagascar, Namibia, South Africa
Western Africa	Benin, Burkina Faso, Cape Verde, Gambia-The, Ghana, Guinea, Guinea-Bissau, Ivory Coast, Liberia, Mali, Mauritania, Niger, Nigeria, Senegal, Sierra Leone, Togo

of data obtained simultaneously from the countries in the sample. Table 3 summarizes the descriptive statistics of the different variables. It indicates small variations, implying unbiased results; however, the correlation matrix contained in Table 4 shows weak interdependencies, suggesting an absence of multicollinearity between the dependent and explanatory variables.

4 Results and Discussion

4.1 *Baseline Results*

Despite concerns about potential endogeneity problems, some interesting elements emerge from the OLS results (Table 5). Indeed, the coefficients of the urban findings are positive and significant in the specification with temperature and rainfall as the dependent variables. In fact, it points to baseline findings that urbanization is highly correlated with the increase in extreme precipitation and rainfall in Africa. Hence, the potential mechanisms known at present can be summarized as follows: (1) urban land cover acts as a heat source with higher temperatures compared to non-urban regions, creating urban heat islands that intensify precipitation through additional instability, and greater moisture transport; (2) urban emissions produce aerosols that increase cloud condensation nuclei; and (3) urban structures work as impediments to atmospheric motion and create eddies with wake diffusion that result in increased precipitation.

The formation and spatiotemporal distribution of precipitation are affected by these mechanisms coupled with topography, regional circulation, and climate. Nonetheless, the contribution of cities to climate change remains noticeable in Africa. More importantly, mitigation and adaptation measures are less promising and conducive to modes of urban development guaranteeing greater viability and

Table 3 Descriptive statistics

Variable	Observation	Mean	Standard deviation	Minimum	Maximum
Temperature	1,560	23.29	3.695	9.30	28.50
Rainfall	1,560	70.65	81.171	0.00	463.20
Urban population growth	1,612	3.65	1.756	−7.18	17.50
Agglomeration population growth	1,054	16.05	9.923	2.83	65.28
Urban population	1,612	41.13	18.907	5.42	94.74
Urban area	120	4,901.94	9285.121	68.24	53,460.31
Foreign development aid	1,519	709,800,000.00	887,500,000.000	−18,410,000.00	11,880,000,000.00
Remittances	1,306	145,600,000.00	298,400,000.000	0.00	3,199,000,000.00
Unemployment	1,530	9.12	7.646	0.30	37.94
Trade openness	1,375	71.21	45.317	0.79	348.00
Agriculture area	1,466	43.64	21.552	2.66	82.67
Domestic investment	798	14.58	7.506	0.00	73.64
Human capital	1,229	93.84	25.939	21.71	156.40
Internet	1,471	8.31	15.206	0.00	86.86
Mobile phone	1,560	67.95	18.587	0.00	204,200,000.00

Table 4 Correlation matrix

Variable	(1)	(2)	(3)	(4)	(5)	(6)	(7)	(8)	(9)	(10)	(11)	(12)	(13)	(14)	(15)
(1) Temperature	1.000														
(2) Rainfall	0.387	1.000													
(3) Urban population growth	0.220	0.409	1.000												
(4) Agglomeration population growth	−0.031	0.404	−0.295	1.000											
(5) Urban population	0.190	0.344	−0.439	0.733	1.000										
(6) Urban area	−0.162	−0.016	−0.607	0.222	0.397	1.000									
(7) Foreign development aid	0.142	0.246	0.223	−0.095	−0.224	0.114	1.000								
(8) Remittances	0.023	0.215	−0.240	0.225	0.483	0.737	−0.029	1.0000							
(9) Unemployment	0.199	0.339	−0.496	0.656	0.769	0.695	−0.152	0.589	1.0000						
(10) Trade openness	0.109	0.660	0.194	0.584	0.506	−0.281	−0.127	−0.081	0.256	1.0000					
(11) Agriculture area	0.003	−0.130	−0.016	−0.197	−0.012	0.220	−0.391	0.319	0.268	0.007	1.0000				
(12) Domestic investment	0.417	0.377	0.250	0.297	0.302	−0.218	0.342	−0.007	0.092	0.474	−0.074	1.0000			
(13) Human capital	−0.397	−0.139	0.101	0.115	0.193	0.097	−0.068	0.124	0.039	0.208	0.256	0.274	1.0000		
(14) Internet	−0.147	−0.070	−0.446	0.269	0.356	0.645	−0.021	0.727	0.413	−0.173	0.031	−0.075	0.184	1.0000	
(15) Mobile phone	−0.177	−0.040	−0.299	0.129	0.159	0.539	0.137	0.589	0.200	−0.207	−0.161	−0.049	0.171	0.920	1.0000

Table 5 Empirical impact of urbanization on climate change in Africa

Variable	OLS estimation			
	Temperature		Rainfall	
Urban population growth	0.285*** (0.0534)	0.136** (0.0972)	4.217*** (1.178)	6.949*** (1.722)
Foreign development aid		−6.39e-10*** (1.94e-10)		−2.88e-10 (3.43e-09)
Remittances		−1.21e-09** (5.54e-10)		4.03e-08*** (9.83e-09)
Unemployment		0.0345 (0.0261)		−0.445 (0.462)
Trade		−0.0107 (0.00728)		0.296** (0.129)
Agriculture area		0.0127* (0.00764)		0.216** (0.135)
Domestic investment		0.0608** (0.0238)		0.391** (0.422)
Constant	22.26*** (0.214)	22.22*** (0.737)	55.39*** (4.729)	13.23 (13.06)
Countries	52	52	52	52
Observations	1.56	597	1.56	597
R-squared	0.018	0.054	0.008	0.065

Note *, **, and *** equate to the significance level at 10, 5, and 1%, respectively; standard deviation robustness to heteroscedasticity is in brackets

greater resilience. However, most of the mechanisms within the international framework on climate change are primarily about governments and do not indicate a clear process that would ensure the participation of local governments and stakeholders. Despite these problems, the current framework, which includes several levels of intervention, offers the possibility of acting at the local level in urban areas. The crux of the problem is constituted by the fact that stakeholders at all levels must work within short deadlines to protect the global level of long-term and far-reaching interests that may seem remote and unpredictable at best.

The results obtained are consistent with those of Yan et al. (2020) and Li et al. (2021), i.e., energy efficiency and renewable energy schemes are part of climate change mitigation efforts. They reduce net GHG emissions and therefore directly combat climatic changes. This is hardly negligible since most of the emissions warming come from cities. The penalties of climate change are already starting to make themselves felt and are likely to continue to affect cities over the coming decades, even under the most favorable scenarios. This adaptation to climate change is important in cities because of the economic and social repercussions of floods or extreme heat waves on populations that are not prepared for them.

Two categories of variables contribute to this result. First, foreign public aid and remittances which mitigate climate change in Africa. Plausible explanations are the implications of aid specifically dedicated to climate change induces the sharing of roles detrimental to the impact of mitigation and development, the loss of sight of the precedence of development goals over adaptation, and the abuse of influence of the climate sector. Also, the promotion of specific climate aid leads to the assurance and demonstration of the climate destination of funding. There is also better tracking of financial contributions and additional, as well as better, tracking and stimulation of progress that contributes to mitigating climate change.

This situation was noticeable following the Copenhagen Accord of 2009, confirmed in the Paris Agreement of 2015, which provided that developed countries would mobilize at least USD 100 billion annually to help developing countries in their efforts to mitigation and adaptation to climate change. But migrant remittances are proving to be an opportunity for adaptation for people facing slow-onset environmental change or disasters. Past experiences of the International Labor Organization have shown that labor migration, when organized in accordance with international labor standards, can play an important role in the development of countries of origin and destination. Second, agricultural area and domestic investment variables which contribute to accentuating climate change in Africa. The explanations relate to the fact that GHG emissions from the agriculture sector represent 24% of global emissions. They are mainly caused by deforestation for agricultural purposes, animal husbandry through enteric fermentation of ruminants and the storage of effluents, and the use of synthetic fertilizers. Africa is responsible for around 15% of global GHG emissions from the agricultural sector. Examples of various forms of agriculture and herding from around Africa are illustrated in Fig. 3.

4.2 *Robustness Analysis*

To test the robustness of the results, we consider an additional measure of urbanization, namely urban population growth. Overall, the results support the positive and statistically significant effect of urbanization on climate change. Temperature and precipitation observed in Africa are statistically influenced by the growth of the urban population. Plausible explanations relate to the fact that today more than half of the world's population lives in urban areas. Even if they only cover 2% of the earth's surface, cities thus concentrate most GHG emissions from the residential, tertiary, industrial, and transport sectors. Population growth will amplify this phenomenon of urban concentration, i.e., cities will expand and with them the artificialization of soils, weakening biodiversity and ecosystems. At the same time, these urban areas are highly vulnerable to climate change by rising sea levels and extreme weather events (i.e., droughts, floods, and storms) that will impact infrastructure and deteriorate access to basic networks and services (i.e., water and energy) (Cirella et al. 2019). The contributory variables to the results are foreign development aid, migrant remittances, unemployment, human capital, internet, and mobile phone usage. The

Fig. 3 Climate impact on different agriculture and herding practices used throughout Africa: (*top left*) low-impact allotments in Nvuze village in Mozambique, (*top right*) tea plantation in Kenya, (*middle*) industrial agriculture in South Africa, (*bottom left*) cattle herding in Uganda, and (bottom right) goat herding in Tanzania. *Source* (*top left*—6 August 2018 and *middle*—25 August 2018) Photographs by Giuseppe T. Cirella; (*top right*) photograph by Antony Trivet on Upsplash, Creative Commons Public Domain, 12 October 2021; (*bottom left*) photograph by Bill Wegener on Upsplash, Creative Commons Public Domain, 27 May October 2021; (*bottom right*) photograph by Rémy Venturini on Upsplash, Creative Commons Public Domain, 18 March 2018

contributory variables are the influencing factors changing temperature and precipitation. One of the direct consequences of the phenomenon of urban sprawl is the increase in GHGs. This for the simple reason that the remoteness of the home from the workplace, which is mainly in the city, requires travel and most often by car (Table 6).

Table 6 GLS regression impact of urban population growth on climate change in Africa

Variable	GLS estimation	
	Temperature	Rainfall
Urban population growth	0.431*** (0.132)	0.201*** (0.0718)
Foreign development aid	−3.31e-10 (2.41e-10)	1.15e-10 (1.15e-10)
Remittances	−1.17e-09* (9.14e-10)	1.49e-09*** (4.37e-10)
Unemployment	0.120*** (0.0316)	0.0169 (0.0152)
Trade	−0.00818 (0.00879)	0.0133*** (0.00430)
Agricultural area	0.00805 (0.00920)	0.0114** (0.00453)
Domestic investment	0.0892*** (0.0305)	−0.0143 (0.0158)
Human capital	−0.0277*** (0.00765)	−0.00836** (0.00383)
Internet	0.0847*** (0.0218)	−0.0166 (0.0105)
Mobile phone	−5.44e-08*** (1.56e-08)	−3.45e-09 (7.36e-09)
Constant	22.18*** (0.964)	1.836*** (0.483)
Countries	52	52
Observations	1530	1530
R-squared	65.42	67.32

Note *, **, and *** equate to the significance level at 10, 5, and 1%, respectively; standard deviation robustness to heteroscedasticity is in brackets

4.3 Sensibility Results

To test the sensibility of the results, three additional measures of urbanization are considered: agglomeration urban population (Table 7), urban area, and urban population. Overall, it appears that the proxies of urbanization considered above have negative and statistically significant effects on temperatures and precipitation in Africa. Although tropical rainforests and ocean spaces play an important role in climate regulation, urban areas are the main contributors to GHG emissions. Large cities concentrate industrial activities, but also the consumption of fossil resources for transport, heating, and air conditioning. The urgency of dealing with the urban fact both in terms of mitigation policies and adaptation capacities because of their impacts on society, although still poorly assessed, are worrying. For populations, the immediate consequences of these emissions are the effects of air pollution on public health. The rapid increase in temperature, linked to the phenomenon of heat islands, also has significant consequences. Despite the lack of urban climate observation systems in African countries, we observe impacts due to the increase in temperature, climate variability, the multiplication of extreme events, and the rise of the oceans. The immediate facts point particularly to landslides and sea level rise. Urban sprawl is a well-known phenomenon in all cities. Territories that were once cut off from urban areas have gradually joined them, eating away square kilometer by square kilometer. Urban areas due to the development of individual houses built between town and country are spread out on the outskirts of agglomerations. This urban growth already poses significant economic, ecological, and human challenges in the short and medium term. It puts the city at the heart of the fight against climate change. From today, but even more so tomorrow, urban areas will concentrate most of the population and economic activity and, consequently, the bulk of the planet's energy consumption and CO_2 emissions.

Indeed, to support the transition of our societies toward more resilient lifestyles in the face of climate change, i.e., the organization and management of cities, their ability to access financial resources and to imagine new sustainable urban models will be decisive (Cirella et al. 2007; Cirella 2023). Therefore, urbanization generates, through the waterproofing of surfaces and the presence of buildings, a local modification of the climate and, more specifically, the phenomenon of the urban heat island (Table 8). This phenomenon is reflected in the increase in temperature in the city at night. During heat waves, this phenomenon can cause discomfort or even excess mortality. Nonetheless, we must not lose sight of the fact that urbanization largely depends on the societal trends of the years to come.

Alternatively, trends in economic growth as well as changes in demographics, lifestyles, and environmental impact may also be interlinked to the COVID-19 crisis and the effects the pandemic has had on African economies since 2020 (Morsy et al. 2019; GIGA 2020; OECD 2020; Anyanwu and Salami 2021). Whatever the scenario, most countries are still expected to urbanize by the end of this century. Also, the organization and management choices of cities, their ability to access financial resources and above all their ability to imagine new sustainable urban models will be decisive in supporting the transition of our societies (Cirella et al. 2007; Cirella and Tao 2008, 2009a, 2009b; Cirella and Zerbe 2014b) toward lifestyles that are more resilient to climate change. Indeed, African countries are rapidly and significantly contributing to climate change. However, given the strong growth of the economies and populations of developing countries, they are also experiencing high levels of CO_2 emissions (Table 9).

The massive entry of digital technologies and telecommunications into the construction and management of the city brings new possibilities to reduce its energy consumption and its GHG emissions (Cirella et al. 2021). The "smart" city indeed offers new possibilities of being a sober city. Smart grids make it possible to make better use of the energy produced by avoiding waste and distributing energy when and where it is needed, while facilitating the use of renewable energies. E-mobility, by providing residents with information on the transport options available nearby

Table 7 Estimate with the share of the population living in urban agglomerations

Variable	GLS estimation	
	Temperature	Rainfall
Agglomeration urban population	−0.119*** (0.0369)	−0.0222** (0.0176)
Foreign development aid	2.24e-10 (2.75e-10)	0.0744** (1.23e-10)
Remittances	−1.12e-09 (1.02e-09)	2.21e-09*** (4.73e-10)
Unemployment	0.222*** (0.0488)	0.0122 (0.0235)
Trade	0.00619 (0.0124)	0.0170*** (0.00581)
Agricultural area	0.00887 (0.0127)	−0.0109* (0.00576)
Domestic investment	0.175*** (0.0350)	0.0508*** (0.0172)
Human capital	−0.0355*** (0.00943)	−0.0131*** (0.00455)
Internet	0.127*** (0.0381)	0.000654 (0.0174)
Mobile phone	−6.98e-08*** (2.19e-08)	−1.37e-08 (9.77e-09)
Constant	22.55*** (0.959)	3.201*** (0.452)
Countries	52	52
Observations	1530	1530
R-squared	62.32	61.74

Note *, **, and *** equate to the significance level at 10, 5, and 1%, respectively; standard deviation robustness to heteroscedasticity is in brackets

(e.g., free parking spaces, public transport timetables, access to carpooling services, etc.), makes public transport more attractive, facilitates the inter-modality of transport, and makes traffic flow more fluid. The ongoing French experiments in this area in Nice and Lyon are very good examples of how digital technologies are reshaping urban environments. These experiments not only showcase the integration of digital technologies but also highlight the importance of performance management and sustainability (Cirella 2020; Cirella et al. 2021; Lewandowski and Cirella 2022). In general, moreover, all urban flows (i.e., energy, waste, and water) find in digital technologies the means to be more efficient by having a lesser impact on natural resources, particularly in energy and, therefore, in GHG emissions. As an example, this enhanced efficiency plays a crucial role in urban sustainable tourism development, as it allows cities to manage their resources more effectively while promoting a

Table 8 Effects of urban area on climate change in Africa

Variable	OLS estimation	
	Temperature	Rainfall
Urban area	−0.000222** (0.000112)	−0.000122*** (4.19e-05)
Foreign development aid	4.28e-10 (1.06e-09)	1.51e-09*** (4.14e-10)
Remittances	4.20e-09 (3.54e-09)	2.86e-09** (1.34e-09)
Unemployment	0.301* (0.179)	0.204*** (0.0648)
Trade	−0.0546* (0.0305)	0.0189* (0.0113)
Agricultural area	−0.00807 (0.0328)	0.00583 (0.0129)
Domestic investment	0.186 (0.160)	−0.198*** (0.0694)
Human capital	−0.0537 (0.0343)	0.00794 (0.0134)
Internet	0.0273 (0.0941)	0.00234 (0.0363)
Mobile phone	−3.85e-08 (5.83e-08)	−5.49e-09 (2.10e-08)
Constant	27.59*** (3.302)	1.378 (1.360)
Countries	52	52
Observations	1530	1530
R-squared	68.31	69.12

Note *, **, and *** equate to the significance level at 10, 5, and 1%, respectively; standard deviation robustness to heteroscedasticity is in brackets

more environmentally friendly and cost-effective experience (Kahraman et al. 2022; Spring and Cirella 2022; Fu et al. 2022).

5 Conclusion

In conclusion, this chapter has analyzed the impact of urbanization on climate change in an African context. Emphasizing that the sustainable urbanization efforts over the last three decades, whose principal beneficiaries are African countries, have theoretically and empirically studied a hitherto unexplored issue. We have opted for OLS and GLS on panel data, whose estimates on a sample of 52 countries over the period 1991–2020 point to the suggestion that urbanization increases climate change in

Table 9 Effects of urban population on climate change in Africa

Variable	GLS estimation	
	Temperature	Rainfall
Urban population	0.0973*** (0.0217)	−0.0873*** (0.0105)
Foreign development aid	−1.30e-10 (2.43e-10)	−9.07e-11 (1.10e-10)
Remittances	−2.02e-09** (9.34e-10)	2.72e-09*** (4.34e-10)
Unemployment	−0.0129 (0.0388)	0.0982*** (0.0177)
Trade	−0.0270*** (0.00881)	0.0194*** (0.00404)
Agricultural area	0.0342*** (0.0106)	−0.0107** (0.00508)
Domestic investment	0.105*** (0.0294)	0.00453 (0.0143)
Human capital	−0.0361*** (0.00798)	0.00378 (0.00374)
Internet	0.0486** (0.0202)	−0.0208** (0.00904)
Mobile phone	−4.58e-08*** (1.54e-08)	−5.07e-09 (6.88e-09)
Constant	21.72*** (0.946)	4.509*** (0.465)
Countries	52	52
Observations	1530	1530
R-squared	61.72	63.96

Note *, **, and *** equate to the significance level at 10, 5, and 1%, respectively; standard deviation robustness to heteroscedasticity is in brackets

Africa. From these results, some recommendations, i.e., outside the comprehensive economic policies needed, can be made in the goal of mitigating climate change in Africa. First, a strengthening of legislation on urbanization should make it possible to establish a sustainable climate change system to control temperature and precipitation levels. Second, governments could consolidate sustainable urbanization strategies to modernize environmental actions. To favor less polluting transport to reduce waste, reuse objects, and recycle them through activities that are less harmful to the environment. Lastly, to implement environmentally sustainable strategies focused on inclusive urbanization.

References

Abebe ST, Dagnew AB, Zeleke VG, Eshetu GZ, Cirella GT (2019) Willingness to pay for watershed management. Resources 8:77. https://doi.org/10.3390/resources8020077

Anyanwu JC, Salami AO (2021) The impact of COVID-19 on African economies: an introduction. Afr Dev Rev 33:S1. https://doi.org/10.1111/1467-8268.12531

Bargaoui SA, Nouri FZ (2017) Dynamic panel data analysis of CO2 emissions driving forces. J Econ Stud Res 2017:1–18. https://doi.org/10.5171/2017.947798

Barrios S, Bertinelli L, Strobl E (2006) Climatic change and rural–urban migration: the case of sub-Saharan Africa. J Urban Econ 60:357–371. https://doi.org/10.1016/J.JUE.2006.04.005

Bradshaw YW (1987) Urbanization and underdevelopment: a global study of modernization, urban bias, and economic dependency. Am Sociol Rev 52:224–239

Cayan DR, Douglas AV (1984) Urban influences on surface temperatures in the Southwestern United States during recent decades - NASA/ADS. J Clim Appl Meteorol 23:1520–1530

Changnon SA (1979) Rainfall changes in summer caused by St. Louis. Science 205:402–404. https://doi.org/10.1126/SCIENCE.205.4404.402

Cirella GT (2020) Sustainable human-nature relations: environmental scholarship, economic evaluation, urban strategies. Springer, Singapore

Cirella GT (2022) Human settlements: urbanization, smart sector development, and future outlook. Springer, Singapore

Cirella GT (2023) Uncertainty shocks in Africa: impact and equilibrium strategies for sound economic and social development. Springer, Cham

Cirella GT, Tao L (2008) Measuring sustainability: an application using the index of sustainable functionality in South East Queensland, Australia. Int J Interdiscip Soc Sci: Annu Rev 3. https://doi.org/10.18848/1833-1882/CGP/v03i08/52680

Cirella GT, Tao L (2009a) An adaptive quantitative method to measure sustainability: an application for the State of Queensland, Australia. Int J Environ Cult Econ Soc Sustain Annu Rev 5:127–139. https://doi.org/10.18848/1832-2077/CGP/v05i01/54563

Cirella GT, Tao L (2009b) The index of sustainable functionality: an application for measuring sustainability. World Acad Sci Eng Technol 3:268–274. https://doi.org/10.5281/zenodo.1330369

Cirella GT, Zerbe S (2014a) Quizzical societies: a closer look at sustainability and principles of unlocking its measurability. Int J Sci Soc 5:29–45. https://doi.org/10.18848/1836-6236/CGP/v05i03/51427

Cirella GT, Zerbe S (2014b) Index of sustainable functionality: application in Urat front banner. In: Cirella GT, Zerbe S (eds) Sustainable water management and wetland restoration in settlements of continental-arid Central Asia. Bozen University Press, Bozen, pp 137–155

Cirella GT, Iyalomhe FO (2019) A determinants of flooding and strategies for mitigation: two-year case study of Benin City. Geosciences (Switzerland) 9. https://doi.org/10.3390/geosciences9030136

Cirella GT, Tao L, Mohamed S (2007) An application of an adaptive quantitative method to measure the sustainability of the gold coast, Australia. J Coast Res SI50:52–56

Cirella GT, Semenzin E, Critto A, Marcomini A (2014) Natural hazard risk assessment and management methodologies review: Europe. In: Linkov I (ed) Sustainable cities and military installations. Springer, Netherlands, Dordrecht, pp 329–358

Cirella GT, Wanjiku S, Paczoski A, Tiruneh S (2020) Human-nature relations: the unwanted filibuster. In: Cirella GT (ed) Sustainable human-nature relations: environmental scholarship, economic evaluation, urban strategies. Springer, Singapore, pp 3–22

Cirella GT, Russo A, Benassi F, Czermanski E, Goncharuk A, Oniszczuk-Jastrząbek A (2021) Energy re-shift for an urbanizing world. Energies 14:5516. https://doi.org/10.3390/en14175516

Colacino M, Rovelli A (2016) The yearly averaged air temperature in Rome from 1782 to 1975. Tellus A: Dyn Meteorol Oceanogr 35 A:389–397. https://doi.org/10.3402/TELLUSA.V35I5.11450

Czermanski E, Cirella GT, Oniszczuk-Jastrząbek A, Pawlowska B, Notteboom T (2021) An energy consumption approach to estimate air emission reductions in container shipping. Energies 14. https://doi.org/10.3390/en14020278

Dash DP, Mallick L (2017) Dynamics of urbanization and temperature increase in middle east-an empirical investigation. Asian Econ Financ Rev 7:486–497. https://doi.org/10.18488/JOURNAL.AEFR/2017.7.5/102.5.486.497

Davidson J (2000) Econometric theory. Blackwell Publishers, Oxford

Du H, Wang D, Wang Y, Zhao X, Qin F, Jiang H, Cai Y (2016) Influences of land cover types, meteorological conditions, anthropogenic heat and urban area on surface urban heat island in the Yangtze River Delta Urban Agglomeration. Sci Total Environ 571:461–470. https://doi.org/10.1016/J.SCITOTENV.2016.07.012

Dumont G-F (2018) Urban demographic transition. Urban Dev Issues 56:13–25. https://doi.org/10.2478/udi-2018-0009

Fu Y, Yang X, Wang T, Supriyadi A, Cirella GT (2022) Spatial pattern characteristics of the financial service industry: evidence from Nanjing, China. Appl Spat Anal Policy 15. https://doi.org/10.1007/s12061-021-09409-z

Gauss CF (1809) Theoria motus corporum coelestium in sectionibus conicis solem ambientium [Translated by C. H. Davis as Theory of the Motion of the Heavenly Bodies Moving about the Sun in Conic Sections, 1857]. Perthes et Besse, Hamburg

Gero AF, Pitman AJ, Narisma GT, Jacobson C, Pielke RA (2006) The impact of land cover change on storms in the Sydney Basin, Australia. Glob Planet Change 54:57–78. https://doi.org/10.1016/J.GLOPLACHA.2006.05.003

GIGA (2020) Africa after the Covid-19 lockdowns: economic impacts and prospects. German Institute for Global and Area Studies. https://www.giga-hamburg.de/en/publications/giga-focus/africa-after-the-covid-19-lockdowns-economic-impacts-and-prospects. Accessed 19 Jul 2022

Goodridge JD (1987) Urban and rural differences in long range temperature trends in California. In: Third conference on climate variations and syrup. On contemporary climate: 1850–2100. American Meteorological Society, Los Angeles, pp 158–159

Hawley AH (1944) Ecology and human ecology. Sociol Forces 22:398–405

Jauregui E, Godinez L, Cruz F (1992) Aspects of heat-island development in Guadalajara, Mexico. Atmos Environ Part B Urban Atmos 26:391–396. https://doi.org/10.1016/0957-1272(92)90014-J

Kahraman C, Orobello C, Cirella GT (2022) Changing dynamics with COVID-19: future outlook. In: Cirella GT (ed) Human settlements: urbanization, smart sector development, and future outlook. Springer, Singapore, pp 235–252

Karaca M, Asian Z (1993) Comparison of urban heat island effects between two largest cities in Turkey: Istanbul and Ankara. In: Abstracts in european geophysical society, XVIII Assembly. FRG, Wiesbaden

Karaca M, Tayanç M, Toros H, Karaca M, Tayanç M, Toros H (1995) Effects of urbanization on climate of Istanbul and Ankara. AtmEn 29:3411–3421. https://doi.org/10.1016/1352-2310(95)00085-D

Kentor J (1981) Structural determinants of peripheral urbanization. Am Sociol Rev 46:201–211. https://doi.org/10.2307/2094979

Kukla G, Gavin J, Karl TR (1986) Urban warming. J Clim Appl Met 25:1265–1270

Kusaka H, Nawata K, Suzuki-Parker A, Takane Y, Furuhashi N (2014) Mechanism of precipitation increase with urbanization in Tokyo as revealed by ensemble climate simulations. J Appl Meteorol Climatol 53:824–839. https://doi.org/10.1175/JAMC-D-13-065.1

Landsberg HE (1981) The urban climate. Academic, New York

Legendre A-M (1805) On least squares [Translated from the French by Professor Henry A Ruger and Professor Helen M Walker, Teachers College, Columbia University, New York City]. University of York. Accessed 10 Jun 2022

Lewandowski R, Cirella GT (2022) Performance management systems: trade-off between implementation and strategy development. Oper Manag Res 16. https://doi.org/10.1007/s12063-022-00305-4

Li W, Chen S, Chen G, Sha W, Luo C, Feng Y, Wen Z, Wang B (2011) Urbanization signatures in strong versus weak precipitation over the pearl river delta metropolitan regions of China. Environ Res Lett 6:034020. https://doi.org/10.1088/1748-9326/6/3/034020

Li Y, Wang W, Chang M, Wang X (2021) Impacts of urbanization on extreme precipitation in the Guangdong-Hong Kong-Macau Greater Bay Area. Urban Clim 38:100904. https://doi.org/10.1016/J.UCLIM.2021.100904

Lin CY, Chen WC, Chang PL, Sheng YF (2011) Impact of the urban heat island effect on precipitation over a complex geographic environment in Northern Taiwan. J Appl Meteorol Climatol 50:339–353. https://doi.org/10.1175/2010JAMC2504.1

Miao S, Chen F, Li Q, Fan S (2011) Impacts of urban processes and urbanization on summer precipitation: a case study of heavy rainfall in Beijing on 1 August 2006. J Appl Meteorol Climatol 50:806–825. https://doi.org/10.1175/2010JAMC2513.1

Moore M, Gould P, Keary BS (2003) Global urbanization and impact on health. Int J Hyg Environ Health 206:269–278. https://doi.org/10.1078/1438-4639-00223

Morsy H, Balma L, Muaksa AN (2019) 'Not a good time': economic impact of COVID-19 in Africa, working paper series 33. African Development Bank, Abidjan

Muntasir M, Syed YS (2018) Effects of urbanization on climate change: evidence from Bangladesh. J Nat Sci Res 8:1–8

OECD (2020) COVID-19 and Africa: socio-economic implications and policy responses. Organisation for EconomicCo-operation and Development. https://www.oecd.org/coronavirus/policy-responses/covid-19-and-africa-socio-economic-implications-and-policy-responses-96e1b282/. Accessed 19 Jul 2022

Oke TR (1979) Review of urban climatology. WMO Tech. Note No. 169, Geneva

Oniszczuk-Jastrząbek A, Czermanski E, Cirella GT (2020) Sustainable supply chain of enterprises: value analysis. Sustainability 12:su12010419. https://doi.org/10.3390/su12010419

Park HS (1986) Features of the heat island in Seoul and its surrounding cities. Atmos Environ 20:1859–1866. https://doi.org/10.1016/0004-6981(86)90326-4

Parnell S, Robinson J (2012) (Re)theorizing cities from the global south: looking beyond neoliberalism. Urban Geogr 33:593–617. https://doi.org/10.2747/0272-3638.33.4.593

Rodrik D (2014) The past, present, and future of economic growth. Challenge 57:5–39. https://doi.org/10.2753/0577-5132570301

Russo A, Cirella GT (2017) Biophilic Cities: Planning for Sustainable and Smart Urban Environments. In: Aijaz R (ed) Smart cities movement in BRICS. Observer Research Foundation and Global Policy Journal, New Delhi, pp 153–159

Russo A, Cirella GT (2020) Urban sustainability: integrating ecology in city design and planning. In: Cirella GT (ed) Sustainable human-nature relations: environmental scholarship, economic evaluation, urban strategies. Springer, Singapore, pp 187–204

Spring C, Cirella GT (2022) Fostering sustainable development: green energy policy in the European Union and the United States. Human settlements: urbanization, smart sector development, and future outlook. Springer, Singapore, pp 101–137

Todaro MP (1969) A model for labor migration and urban unemployment in less developed countries. Am Econ Rev 59:138–148

UN-Habitat (2018) The state of African cities 2018: The geography of African investment. United Nations Human Settlements Programme, Geneva

Veremchuk LV, Yankova VI, Vitkina TI, Nazarenko AV, Golokhvast KS (2016) Urban air pollution, climate and its impact on asthma morbidity. Asian Pac J Trop Biomed 6:76–79. https://doi.org/10.1016/j.apjtb.2015.10.001

Wang Y, Wan Q, Meng W, Liao F, Tan H, Zhang R (2011) Long-term impacts of aerosols on precipitation and lightning over the pearl river delta megacity area in China. Atmos Chem Phys 11:12421–12436. https://doi.org/10.5194/ACP-11-12421-2011

Wang X, Liao J, Zhang J, Shen C, Chen W, Xia B, Wang T (2014) A numeric study of regional climate change induced by urban expansion in the pearl river delta, China. J Appl Meteorol Climatol 53:346–362. https://doi.org/10.1175/JAMC-D-13-054.1

Wen J, Chen JI, Lin W, Jiang B, Xu S, Lan J (2020) Impacts of anthropogenic heat flux and urban land-use change on frontal rainfall near coastal regions: a case study of a rainstorm over the pearl river delta, South China. J Appl Meteorol Climatol 59:363–379. https://doi.org/10.1175/JAMC-D-18-0296.1

Wigley TML, Farmer G (1982) Climate of the eastern mediterranean and near east. In: Bintliff JL, Van Zeist W (eds) Palaeoclimates, palaeoenvironments and human communities in the eastern mediterranean region in later prehistory. BAR International Series, London, pp 1–37

Wikipedia (2022) Geopolitical zones of Nigeria. https://en.wikipedia.org/wiki/Geopolitical_zones_of_Nigeria. Accessed 15 Jul 2022

Wooldridge JM (2010) Econometric analysis of cross section and panel data. MIT Press, Chicago

World Bank (2020) World development indicators 2020. WDI World Bank. https://datatopics.worldbank.org/world-development-indicators/. Accessed 18 Jul 2022

World Bank (2021) Climate change knowledge portal. https://climateknowledgeportal.worldbank.org/. Accessed 18 Jul 2022

Xiao H, Yin Y, Jin L, Chen Q, Chen J (2014) Simulation of aerosol effects on orographic clouds and precipitation using WRF model with a detailed bin microphysics scheme. Atmos Sci Lett 15:134–139. https://doi.org/10.1002/ASL2.480

Yagüe C, Zurita E, Martinez A (1991) Statistical analysis of the Madrid urban heat island. Atmos Environ Part B Urban Atmos 25:327–332. https://doi.org/10.1016/0957-1272(91)90004-X

Yan M, Chan JCL, Zhao K, Yan M, Chan JCL, Zhao K (2020) Impacts of urbanization on the precipitation characteristics in Guangdong Province, China. Adv Atmos Sci 37:696–706. https://doi.org/10.1007/S00376-020-9218-3

Yang L, Smith JA, Baeck ML, Bou-Zeid E, Jessup SM, Tian F, Hu H (2014) Impact of urbanization on heavy convective precipitation under strong large-scale forcing: a case study over the Milwaukee Lake Michigan Region. J Hydrometeorol 15:261–278. https://doi.org/10.1175/JHM-D-13-020.1

Yu M, Liu Y, Miao S (2020) Impact of urbanization on rainfall of different strengths in the Beijing area. Theoret Appl Climatol 139:1097–1110. https://doi.org/10.1007/S00704-019-03035-Z/TABLES/3

Zhang L, Tan PY, Diehl JA (2017a) A conceptual framework for studying urban green spaces effects on health. J Urban Ecol 3. https://doi.org/10.1093/jue/jux015

Zhang Q, Zheng Y, Singh VP, Luo M, Xie Z (2017b) Summer extreme precipitation in eastern China: mechanisms and impacts. J Geophys Res: Atmos 122:2766–2778. https://doi.org/10.1002/2016JD025913

Zhang Y, Miao S, Dai Y, Bornstein R (2017c) Numerical simulation of urban land surface effects on summer convective rainfall under different UHI intensity in Beijing. J Geophys Res: Atmos 122:7851–7868. https://doi.org/10.1002/2017JD026614

Glossary

Bike Sharing A system of the so-called city bike, i.e., a transport service in which a specific fleet of bicycles is made available to residents for short-term use (sharing)—on preferential payment terms for users in the public model (subsidized by the city) or on the initiative of a private operator, then on a commercial basis (Mobilne Miasto 2022).

Climate Action Stepped-up efforts to reduce greenhouse gas emissions and strengthen resilience and adaptive capacity to climate-induced impacts, including: climate-related hazards in all countries; integrating climate change measures into national policies, strategies, and planning; and improving education, awareness-raising, and human and institutional capacity with respect to climate change mitigation, adaptation, impact reduction, and early warning (UNDP 2015).

Climate Change Any change in the state of the climate that can be identified by changes in the mean and variability of its properties, and that persists for an extended period, typically decades or longer (IPCC 2007).

Cooperatives An autonomous association of persons united voluntarily to meet their common economic, social, and cultural needs and aspirations through a jointly owned and democratically controlled enterprise. Another definition is a people-centered enterprises owned, controlled, and run by and for their members to realize their common economic, social, and cultural needs and aspirations (ICA 2023).

Development The economic, social, and institutional evolution of national states (De Blij and Murphy 2003). Can be interpreted to mean the process of becoming larger, more mature, and better organized—often in terms of economic criteria (Norton 1998).

Disaster Management A systematic process of using the body of policy, administrative decisions, organization, and operational skills and capacities to reduce the risk of natural or man-made disasters and their adverse effects. Disaster management can be realized at various levels—sub-national (i.e., local, regional), national, and international, and at various times—before, during, and after a disaster (Sammut-Bonnici 2015).

G. T. Cirella and B. Dahiya (eds.), *City Responses to Disruptions in 2020*, Advances in 21st Century Human Settlements, https://doi.org/10.1007/978-981-99-7988-2

Econometrics A method that uses economic theory, mathematics, and statistical inference to quantify economic phenomena. In other words, it turns theoretical economic models into useful tools for economic policymaking. The objective of econometrics is to convert qualitative statements (such as "the relationship between two or more variables is positive") into quantitative statements (such as "consumption expenditure increases by 95 cents for every one dollar increase in disposable income"). The main tool of econometrics is the linear multiple regression model, which provides a formal approach to estimating how a change in one economic variable, the explanatory variable, affects the variable being explained, the dependent variable—taking into account the impact of all the other determinants of the dependent variable (Ouliaris 2011).

Ecosystem Services The direct and indirect contributions of ecosystems to human well-being. There are four types of ecosystem services: provisioning, regulating, supporting, and cultural (TEEB 2019).

Emergency Services The emergency services are the public organizations whose job is to take quick action to deal with emergencies when they occur, especially the fire brigade, the police, and the ambulance service (Collins Dictionary 2023).

Energy Consumption The total amount of energy required for a given process and is measured in kilowatt hours (kWh). This includes the use of electricity, gas, diesel, oil, and biomass (Repsol 2023).

Enterprise Another word for a for-profit business or company, but it is most often associated with entrepreneurial ventures (Shopify Staff 2022).

Environmental Footprint Also known as the ecological footprint, measures the impact of people or organizations on the environment, whether it is positive or negative. In the case of organizations, it is defined as a multi-criteria measure of the environmental performance of their activity when providing goods or services, taking into account the entire life cycle (Aplanet 2022).

Environmental Pollution Defined as the contamination of the physical and biological components of the earth or atmosphere system to such an extent that normal environmental processes are adversely affected (Muralikrishna and Manickam 2017).

Green City A loose association of a city focused on sustainability, striving to lessen its environmental impact by reducing waste, expanding recycling, lowering emissions, increasing housing density while expanding open blue and green spaces, and encouraging the development of sustainable local businesses (E/The Environmental Magazine 2009).

Green Energy Green energy is any energy type that is generated from natural resources, such as sunlight, wind, or water. It often comes from renewable energy sources (see definition below) although there are some differences between renewable and green energy. The key with these energy resources is that they do not harm the environment through factors such as releasing greenhouse gases into the atmosphere (TWI 2021).

Human Settlement A form of human habitation which ranges from a single dwelling to large urban center. It is a process of opening up and settling of a previously uninhabited area by people. The study of human settlement is basic to human geography as it examines the forms of settlement in any particular region reflective of the human relationship with the environment (Josh 2015).

Informal Sector Also known as the informal economy, comprises of activities that have market value and would add to tax revenue and gross domestic product if they were recorded, is a globally widespread phenomenon. According to the International Labor Organization, about 2 billion workers, or 60% of the world's employed population ages 15 and older, spend at least part of their time in the informal sector. The size of the informal sector slowly decreases as economies develop, but with wide variations across regions and countries. Today, the informal sector still accounts for about a third of low- and middle-income countries' economic activity—15% in advanced economies (IMF 2023).

Legislation The preparing and enacting of laws by local, state, or national legislatures. In other contexts, it is sometimes used to apply to municipal ordinances and to the rules and regulations of administrative agencies passed in the exercise of delegated legislative functions (Britannica 2021).

Literature Review A piece of academic writing demonstrating knowledge and understanding of the academic literature on a specific topic placed in context. A literature review also includes a critical evaluation of the material; this is why it is called a literature review rather than a literature report. It is a process of reviewing the literature, as well as a form of writing (Institute for Academic Development 2023).

Methodology Is a theory of producing knowledge through research and provides a rationale for the way a researcher proceeds. It is the philosophical underpinning of a given research practice. Methodology is the discussion of the theory upon which research is based and indicates how the experimentation is performed (Zeegers et al. 2015).

Natural Disasters Catastrophic events with atmospheric, geological, and hydrological origins (e.g., droughts, earthquakes, floods, hurricanes, landslides) that can cause fatalities, property damage, and social environmental disruption (Xu et al. 2016).

Panel Data Analysis Also known as longitudinal data analysis or cross-sectional time-series analysis, is a statistical method used in social sciences, economics, and other fields to examine data collected over multiple time periods and from multiple individuals or entities. It combines elements of both cross-sectional data (i.e., data collected at a single point in time from different subjects) and time-series data (i.e., data collected over multiple time points from the same subjects). In panel data analysis, researchers study how variables change over time within individual subjects or entities and across different subjects. Common techniques in panel data analysis include fixed effects models, random effects models, and various forms of regression analysis, which help researchers understand relationships, trends, and patterns in the data over time (Xu et al. 2008; Fitrianto and Musakkal 2016; Ioan et al. 2020; Sarafidis and Wansbeek 2021).

Personal Protective Equipment (PPE) This refers to specialized clothing, gear, or equipment designed to protect individuals from various hazards, such as physical, chemical, biological, or environmental, in the workplace or other settings. The primary purpose of PPE is to safeguard the wearer's health and safety by minimizing exposure to potential risks. Common examples of PPE include: safety helmets and hard hats to protect the head from falling objects or impacts, gloves to protect the hands from chemicals, sharp objects, or extreme temperatures, and safety vests or full-body suits for visibility and protection against various workplace hazards (Steuart et al. 2020; Kim et al. 2021; Galanis et al. 2021).

Plastic Waste This refers to discarded or unwanted plastic materials, such as bottles, bags, containers, packaging, and other items made from synthetic polymers like polyethylene, polypropylene, and polyvinyl chloride. Plastic waste can take various forms and can be found in diverse settings, including households, businesses, landfills, oceans, and natural environments. Plastic waste is a significant environmental concern due to its durability and persistence in the environment. Efforts to address plastic waste include recycling, reducing plastic consumption, promoting responsible disposal, and developing alternatives to single-use plastics to mitigate its environmental impact (Evode et al. 2021; Kibria et al. 2023).

Policy A law, regulation, procedure, administrative action, incentive, or voluntary practice of governments and other institutions (CDC 2021).

Public Spaces These spaces refer to areas or locations that are accessible and open to all members of the community or the general public. These areas can be owned and maintained by the government, private entities, or non-profit organizations and are designed for various purposes, including recreation, social interaction, cultural events, and civic activities. Examples of public spaces include parks, plazas, sidewalks, libraries, community centers, and public transportation hubs (Siláči and Vitková 2017; Li et al. 2022).

Renewables Energy that is derived from sources or processes that are constantly replenished. These sources of energy include solar energy, wind energy, geothermal energy, and hydroelectric power (TWI 2021).

Resilience This is the ability of a system, community, or individual to withstand, adapt to, and recover from various stressors, shocks, or disturbances while maintaining its essential functions, structures, and identity. Resilience can be applied to different contexts, such as environmental resilience (the ability of ecosystems to withstand and recover from environmental changes), economic resilience (the capacity of an economy to absorb and bounce back from economic shocks), and personal resilience (an individual's ability to cope with adversity and bounce back from challenges) (Herrman et al. 2011; Southwick et al. 2014).

Resilient Communities These are communities that are well-prepared to face and recover from adverse events and shocks, such as natural disasters, economic crises, or social disruptions. These communities often have robust infrastructure, effective emergency response systems, strong social networks, and the capacity to adapt to changing circumstances. Resilient communities prioritize risk reduction, disaster preparedness, and long-term sustainability to ensure the well-being of their residents in the face of challenges (Carta et al. 2021; Fan and Lyu 2021; Korosteleva and Petrova 2022).

Shocks In the context of resilience and community planning, shocks refer to sudden and disruptive events or circumstances that can have a significant negative impact on a system or community. These shocks can be diverse and include natural disasters (e.g., earthquakes, hurricanes), economic crises (e.g., recessions, market crashes), public health emergencies (e.g., pandemics), and social disruptions (e.g., riots, civil unrest). Shocks are typically unexpected or unpredictable and can challenge the stability and functioning of a system or community (Cochrane 1994; Bazzi and Blattman 2014; Lafrogne-Joussier et al. 2023).

Survey A research method used for collecting data from a predefined group of respondents to gain information and insights into various topics of interest (QuestionPro 2020).

Sustainability Development that meets the needs of the present without compromising the ability of future generations to meet their own needs (Brundtland 1987).

Unionization This refers to the process of workers coming together to form labor unions, which are organizations designed to represent and advocate for the collective interests and rights of employees in a particular industry, company, or sector. Through unionization, workers seek to negotiate better working conditions, wages, benefits, and workplace policies with their employers. Unions can also engage in collective bargaining on behalf of their members and may participate in various labor-related activities such as strikes and protests to achieve their goals (ScienceDirect Topics 2023).

Urbanization This is the demographic and social process by which an increasing proportion of a population moves from rural areas to cities and towns, leading to the growth and expansion of urban areas. This phenomenon involves the conversion of rural land into urban land, resulting in higher population densities and greater economic, social, and cultural integration within urban centers. Urbanization is often associated with various changes in lifestyle, infrastructure, and economic activities (Ritchie and Roser 2018; Kuddus et al. 2020).

Urban Rewilding This is a concept and practice that involves restoring and reintroducing elements of nature and biodiversity in urban environments. It aims to enhance the ecological health and resilience of cities by reintroducing native plant species, creating green spaces, and promoting wildlife habitats within urban areas. Urban rewilding initiatives often seek to improve air and water quality, mitigate the urban heat island effect, and provide opportunities for urban residents to connect with nature (Song 2023; Moxon et al. 2023).

Urban Sustainability This refers to the effort to create and maintain urban environments that meet the needs of the present generation without compromising the ability of future generations to meet their own needs. It involves the responsible and efficient use of resources, reducing environmental impacts, promoting social equity, and ensuring economic prosperity within urban areas. Key aspects of urban sustainability include sustainable urban planning, renewable energy adoption, waste reduction, public transportation, green building practices, and community engagement (Spiliotopoulou and Roseland 2020; Sharifi 2021).

Volunteers They are individuals who willingly offer their time, skills, and services to assist others or contribute to various causes, organizations, or activities without expecting monetary compensation. Volunteers typically engage in a wide range of activities, such as providing support to charitable organizations, assisting in disaster relief efforts, participating in community service projects, mentoring, tutoring, or offering their expertise in areas like healthcare, education, or social services. Volunteerism plays a crucial role in addressing social, environmental, and humanitarian issues and contributes to the betterment of society as a whole (Yeung et al. 2017; Help Guide 2023).

Well-being The degree to which the needs and wants of a society are satisfied (Norton 1998). An individual's perception of their position in life in the context of the culture and value systems in which they live and in relation to their goals, expectations, standards, and concerns. It is a broad ranging concept affected in a complex way by the person's physical health, psychological state, personal beliefs, social relationships, and their relationship to salient features of their environment (WHO 2014).

References

Aplanet (2022) What is the environmental footprint and how is it measured? Aplanet. https://aplanet.org/resources/what-is-the-environmental-footprint-and-how-is-it-measured/. Accessed 12 Sep 2023

Bazzi S, Blattman C (2014) Economic shocks and conflict: evidence from commodity prices. Am Econ J Macroecon 6:1–38. https://doi.org/10.1257/mac.6.4.1

Britannica (2021) legislation|Definition. Britannica dictionary. https://www.britannica.com/topic/legislation-politics. Accessed 24 Mar 2021

Brundtland GH (1987) Our common future: the world commission on environment and development. Oxford University Press, Oxford

Carta S, Pintacuda L, Owen IW, Turchi T (2021) Resilient communities: a novel workflow. Front Built Environ 7

CDC (2021) Definition of policy. Centers for Disease Control and Prevention. https://www.cdc.gov/policy/analysis/process/definition.html. Accessed 24 Mar 2021

Cochrane JH (1994) Shocks. Carn-Roch Conf Ser Public Policy 41:295–364. https://doi.org/10.1016/0167-2231(94)00024-7

Collins Dictionary (2023) Emergency services definition and meaning. Collins Dictionary. https://www.collinsdictionary.com/dictionary/english/emergency-services. Accessed 12 Sep 2023

De Blij HJ, Murphy AB (2003) Human geography: Culture, society, and space. Wiley, Hoboken, NJ

E/The Environmental Magazine (2009) What are "green cities"? Phys.org. https://phys.org/news/2009-03-green-cities.html. Accessed 24 Mar 2021

Evode N, Qamar SA, Bilal M, Barceló D, Iqbal HMN (2021) Plastic waste and its management strategies for environmental sustainability. Case Stud Chem Environ Eng 4:100142. https://doi.org/10.1016/j.cscee.2021.100142

Fan Y, Lyu X (2021) Exploring two decades of research in community resilience: a content analysis across the international literature. PRBM 14:1643–1654. https://doi.org/10.2147/PRBM.S329829

Fitrianto A, Musakkal NFK (2016) Panel data analysis for sabah construction industries: choosing the best model. Procedia Econ Financ 35:241–248. https://doi.org/10.1016/S2212-5671(16)00030-7

Galanis P, Vraka I, Fragkou D, Bilali A, Kaitelidou D (2021) Impact of personal protective equipment use on health care workers' physical health during the COVID-19 pandemic: a systematic review and meta-analysis. Am J Infect Control 49:1305–1315. https://doi.org/10.1016/j.ajic.2021.04.084

G. T. Cirella and B. Dahiya (eds.), *City Responses to Disruptions in 2020*, Advances in 21st Century Human Settlements, https://doi.org/10.1007/978-981-99-7988-2

Help Guide (2023) Volunteering and its surprising benefits. Help Guide. https://www.helpguide.org/articles/healthy-living/volunteering-and-its-surprising-benefits.htm. Accessed 13 Sep 2023

Herrman H, Stewart DE, Diaz-Granados N, Berger EL, Jackson B, Yuen T (2011) What is resilience? Can J Psychiatr 56:258–265. https://doi.org/10.1177/070674371105600504

ICA (2023) What is a cooperative? International Cooperative Alliance. https://www.ica.coop/en/cooperatives/what-is-a-cooperative. Accessed 12 Sep 2023

IMF (2023) What is the informal economy? IMF. https://www.imf.org/en/Publications/fandd/issues/2020/12/what-is-the-informal-economy-basics. Accessed 12 Sep 2023

Institute for Academic Development (2023) Literature review. The University of Edinburgh. https://www.ed.ac.uk/institute-academic-development/study-hub/learning-resources/literature-review. Accessed 12 Sep 2023

Ioan B, Malar Kumaran R, Larissa B, Anca N, Lucian G, Gheorghe F, Horia T, Ioan B, Mircea-Iosif R (2020) A panel data analysis on sustainable economic growth in India, Brazil, and Romania. J Risk Financ Manag 13:170. https://doi.org/10.3390/jrfm13080170

IPCC (2007) Climate change 2007: the physical science basis. https://www.ipcc.ch/report/ar4/wg1/. Accessed 30 Oct 2019

Josh J (2015) Human settlement. https://www.jagranjosh.com/general-knowledge/human-settlement-1448446636-1. Accessed 30 Oct 2019

Kibria MdG, Masuk NI, Safayet R, Nguyen HQ, Mourshed M (2023) Plastic waste: challenges and opportunities to mitigate pollution and effective management. Int J Environ Res 17:20. https://doi.org/10.1007/s41742-023-00507-z

Kim H, Hegde S, LaFiura C, Raghavan M, Sun N, Cheng S, Rebholz CM, Seidelmann SB (2021) Access to personal protective equipment in exposed healthcare workers and COVID-19 illness, severity, symptoms and duration: a population-based case-control study in six countries. BMJ Glob Health 6:e004611. https://doi.org/10.1136/bmjgh-2020-004611

Korosteleva EA, Petrova I (2022) What makes communities resilient in times of complexity and change? Camb Rev Int Aff 35:137–157. https://doi.org/10.1080/09557571.2021.2024145

Kuddus MA, Tynan E, McBryde E (2020) Urbanization: a problem for the rich and the poor? Public Health Rev 41:1. https://doi.org/10.1186/s40985-019-0116-0

Lafrogne-Joussier R, Martin J, Mejean I (2023) Supply shocks in supply chains: evidence from the early lockdown in China. IMF Econ Rev 71:170–215. https://doi.org/10.1057/s41308-022-00166-8

Li J, Dang A, Song Y (2022) Defining the ideal public space: a perspective from the publicness. J Urban Manag 11:479–487. https://doi.org/10.1016/j.jum.2022.08.005

Mobilne Miasto (2022) Bike sharing Czyli Rower Miejski. Mobilne Miasto. https://mobilne-miasto.org/bike-sharing-czyli-rower-miejski/. Accessed 12 Sep 2023

Moxon S, Webb J, Semertzi A, Samangooei M (2023) Wild ways: a scoping review to understand urban-rewilding behaviour in relation to adaptations to private gardens. Cities & Health 7:888–902. https://doi.org/10.1080/23748834.2023.2218016

Muralikrishna IV, Manickam V (2017) Introduction. In: Environmental management. Elsevier, pp 1–4

Norton W (1998) Human geography. Oxford University Press

Ouliaris S (2011) Finance and development. Finance and Development. https://www.imf.org/external/pubs/ft/fandd/2011/12/basics.htm. Accessed 12 Sep 2023

QuestionPro (2020) What is a survey|Definition, characteristics, examples of a good survey. https://www.questionpro.com/blog/surveys/. Accessed 24 Mar 2021

Repsol (2023) What is energy consumption and why is it important? Repsol. https://www.repsol.com/en/energy-and-the-future/future-of-the-world/energy-consumption/index.cshtml. Accessed 12 Sep 2023

Ritchie H, Roser M (2018) Urbanization. Our world in data

Sammut-Bonnici T (2015) Complexity theory. Wiley encyclopedia of management. Wiley, Chichester, pp 1–2

Sarafidis V, Wansbeek T (2021) Celebrating 40 years of panel data analysis: past, present and future. J Econom 220:215–226. https://doi.org/10.1016/j.jeconom.2020.06.001
ScienceDirect Topics (2023) Unionization: an overview. Science Direct Topics. https://www.sciencedirect.com/topics/earth-and-planetary-sciences/unionization. Accessed 13 Sep 2023
Sharifi A (2021) Urban sustainability assessment: an overview and bibliometric analysis. Ecol Ind 121:107102. https://doi.org/10.1016/j.ecolind.2020.107102
Shopify Staff (2022) What is an enterprise? Definition and guide. Shopify. https://www.shopify.com/blog/what-is-an-enterprise. Accessed 12 Sep 2023
Siláči I, Vitková L (2017) Public spaces as the reflection of society and its culture. IOP Conf Ser: Mater Sci Eng 245:042009. https://doi.org/10.1088/1757-899X/245/4/042009
Song F (2023) Rewilding cities: Why it's needed and how you can help. Mossy Earth. https://www.mossy.earth/rewilding-knowledge/rewilding-cities. Accessed 13 Sep 2023
Southwick SM, Bonanno GA, Masten AS, Panter-Brick C, Yehuda R (2014) Resilience definitions, theory, and challenges: interdisciplinary perspectives. Eur J Psychotraumatol 5. https://doi.org/10.3402/ejpt.v5.25338
Spiliotopoulou M, Roseland M (2020) Urban sustainability: from theory influences to practical agendas. Sustainability 12:7245. https://doi.org/10.3390/su12187245
Steuart R, Huang FS, Schaffzin JK, Thomson J (2020) Finding the value in personal protective equipment for hospitalized patients during a pandemic and beyond. J Hosp Med 15:295–298. https://doi.org/10.12788/jhm.3429
TEEB (2019) Ecosystem services. The Economics of Ecosystems and Biodiversity. http://www.teebweb.org/resources/ecosystem-services/. Accessed 30 Oct 2019
TWI (2021) What is green energy? https://www.twi-global.com/technical-knowledge/faqs/what-is-green-energy. Accessed 24 Mar 2021
UNDP (2015) Goal 13: climate action. United Nations Sustainable Development Goals. https://www.sdfinance.undp.org/content/sdfinance/en/home/sdg/goal-13--climate-action.html. Accessed 24 Mar 2021
WHO (2014) WHOQOL: measuring quality of life. World Health Organization. https://www.who.int/healthinfo/survey/whoqol-qualityoflife/en/. Accessed 30 Oct 2019
Xu D, Lee SH, Eom T (2008) Introduction to panel data analysis. In: MIller G, Yang K (eds) Handbook of research methods in public administration, 2nd edn. Routledge, New York, p 990
Xu J, Wang Z, Shen F, Ouyang C, Tu Y (2016) Natural disasters and social conflict: a systematic literature review. Int J Disas Risk Reduct 17:38–48. https://doi.org/10.1016/j.ijdrr.2016.04.001
Yeung JWK, Zhang Z, Kim TY (2017) Volunteering and health benefits in general adults: cumulative effects and forms. BMC Public Health 18:8. https://doi.org/10.1186/s12889-017-4561-8
Zeegers M, Barron D, Zeegers M, Barron D (2015) Milestone 5: methodology. Milestone moments in getting your PhD in qualitative research, pp 61–74. https://doi.org/10.1016/B978-0-08-100231-5.00005-5

Index

G. T. Cirella and B. Dahiya (eds.), *City Responses to Disruptions in 2020*, Advances in 21st Century Human Settlements, https://doi.org/10.1007/978-981-99-7988-2

GPSR Compliance

The European Union's (EU) General Product Safety Regulation (GPSR) is a set of rules that requires consumer products to be safe and our obligations to ensure this.

If you have any concerns about our products, you can contact us on ProductSafety@springernature.com

In case Publisher is established outside the EU, the EU authorized representative is:

Springer Nature Customer Service Center GmbH
Europaplatz 3
69115 Heidelberg, Germany

Batch number: 10756376

Printed by Printforce, the Netherlands